Involving Anthroponomy in the Anthropocene

This book introduces the idea of anthroponomy – the organization of humankind to support autonomous life – as a response to the problems of today's purported "Anthropocene" age. It argues for a specific form of accountability for the redressing of planetary-scaled environmental problems.

The concept of anthroponomy helps confront geopolitical history shaped by the social processes of capitalism, colonialism, and industrialism, which have resulted in our planetary situation. *Involving Anthroponomy in the Anthropocene: On Decoloniality* explores how mobilizing our engagement with the politics of our planetary situation can come from moral relations. This book focuses on the anti-imperial work of addressing unfinished decolonization, and hence involves the "decolonial" work of cracking open the common sense of the world that supports ongoing colonization. "Coloniality" is the name for this common sense, and the discourse of the "Anthropocene" supports it. A consistent anti-imperial and anti-capitalist politics, one committed to equality and autonomy, will problematize the Anthropocene through decoloniality. Sometimes the way forward is the way backward.

Written in a novel style that demonstrates – not simply theorizes – moral relatedness, this book makes a valuable contribution to the fields of Anthropocene studies, environmental studies, decolonial studies, and social philosophy.

Jeremy Bendik-Keymer holds the Beamer-Schneider Professorship in Ethics at Case Western Reserve University. He authored *The Ecological Life* and two works in literary philosophy, *Solar Calendar, and Other Ways of Marking Time* and *The Wind*. He co-edited *Ethical Adaptation to Climate Change*.

Routledge Research in the Anthropocene
Series Editors: Jamie Lorimer and Kathryn Yusoff

The *Routledge Research in the Anthropocene Series* offers the first forum for original and innovative research on the epoch and events of the Anthropocene. Titles within the series are empirically and/or theoretically informed and explore a range of dynamic, captivating and highly relevant topics, drawing across the humanities and social sciences in an avowedly interdisciplinary perspective. This series will encourage new theoretical perspectives and highlight ground-breaking interdisciplinary research that reflects the dynamism and vibrancy of current work in this field. The series is aimed at upper-level undergraduates, researchers and research students as well as academics and policy-makers.

Hope and Grief in the Anthropocene
Re-conceptualising Human–Nature Relations
Lesley Head

Releasing the Commons
Edited by Ash Amin and Philip Howell

Climate Change Ethics and the Non-Human World
Edited by Brian G. Henning and Zack Walsh

Involving Anthroponomy in the Anthropocene
On Decoloniality
Jeremy Bendik-Keymer

For more information about this series, please visit www.routledge.com/Routledge-Research-in-the-Anthropocene/book-series/RRA01

Involving Anthroponomy in the Anthropocene

On Decoloniality

Jeremy Bendik-Keymer

LONDON AND NEW YORK

First published 2020
by Routledge
2 Park Square, Milton Park, Abingdon, Oxon OX14 4RN

and by Routledge
52 Vanderbilt Avenue, New York, NY 10017

Routledge is an imprint of the Taylor & Francis Group, an informa business

British Library Cataloguing-in-Publication Data
A catalogue record for this book is available from the British Library

Library of Congress Cataloging-in-Publication Data
A catalog record for this book has been requested

ISBN: 978-1-138-54953-1 (hbk)
ISBN: 978-1-351-00023-9 (ebk)

Typeset in Times New Roman
by Apex CoVantage, LLC

Figure 0.1 Shaker Heights, Ohio, summer 2015

In memoriam, Esther Ann Bendik (1939–2018).

We will wait half a century for our talk to sing.

Contents

Figures

Acknowledgments

I want to acknowledge the older nations that lived in the "Northwest Territory" region of the Midwest before settler colonialism. Just two decades after the Treaty of Greenville (1795), made under threat of the U.S. military, a small part of this land became Cleveland, Ohio. The older nations include:

The Chippewa
The Kaskaskia
The Kickapoo
The Lenape
The Ottawa
The Piankashaw
The Potawatomi
The Shawnee
The Wea, and
The Wyandot Miami.

These nations had other names as well. The Treaty of Greenville was not honored by settlers or by the United States of America.

*

I discussed the title of this book with my mother, Esther Bendik, in November 2017 while I was in Sydney, Australia. The title reminds me of my mother's attention to words. She was struggling with cancer. It took her life less than a year later. This book is indebted to the way she approached her relationships, including her relationship to herself as a person who had philosophical room to grow.

There were other good things about my time in Sydney. They included hearing, for the third time that fall, Kyle Powys Whyte (Citizen Potawatomi Nation) shake up the categories of environmental justice and of social and political thought. This book is indebted to his energy, scholarship, and activism. I want also to thank Susan Dominguez and Bruce Kafer (Oglala Lakota Nation) who joined in around Kyle's visit to Cleveland earlier that fall. Finally, Kyle made me aware of Julia D. Gibson, whom I later met at the International Society for Environmental Ethics 2019 annual meeting.

This project began with a post on Zev Trachtenberg's blog, *Inhabiting the Anthropocene*. I want to thank Faye Leerink and her team-members at Routledge, Ruth Anderson, and Nonita Saha; Cathy Hurren at Taylor & Francis; Ganesh Pawan Kumar Agoor at Apex, and Stacy Lavin; Sarah Jammal for assisting with research and for their indexing; Steven Vogel for offering early feedback; and John Levy Barnard, Sarah Gridley, and Allen Thompson for earlier discussions. Funds for some work were provided by the Beamer-Schneider Professorship in Ethics, with thanks to Renée Holland-Golphin for administrative support.

I am grateful to Todd Cleveland, Ananya Dasgupta, and Laura Hengehold for initially discussing colonialism; David Beach for sharing a history of the settlement of the Western Reserve and for lending me his books on bioregionalism; Cullin Brown, Judy Twedt, Jason Walsh, and Peter Whitehouse for providing comments on the first draft – Cullin specifically for his defense of autonomy as intrinsically anthroponomous; the *Changing the Politics of Earth* workshop of the 2019 Cleveland Humanities Festival, with support from the Baker Nord Center for the Humanities, Cuyahoga Arts and Culture, and the Cleveland Botanical Gardens; and Stephen M. Rich for discussions about identity and race. I am grateful to Joel Wainwright for extensive commentary and critique and to Rupert Read for a conversation in Russell Square, London. Finally, I am grateful to Julia D. Gibson for writing her response to this book in her voice.

While writing this book, I have been fortunate to live with my family and spend time with my friends. I am thankful to Lars Helge-Strand for hospitality one summer. I am grateful for the time I have had with my father, David Keymer, swimming and talking at the pool. To Misty, especially, I am grateful for the time we have had in the years surrounding this book with Atlas, Phoebe, and now Emet in our world. Around us all, too, there is the more than human world for which I am most thankful.

Finally, to anyone who has protested the order of things for the sake of a saner world to come, being honest about the dire future we all face and about the wrongness of what is often called "civilization" now, my entire family thanks you. It is time for this world to change.

Jeremy Bendik-Keymer
March 30th, 2020

Preface: On the essay form

Some years ago, because I am anxious and overwhelmed by the seemingly inexorable power by which global capitalism and the existing world system of states accelerate the slide into mass extinction and extreme global warming, I fantasized about a political movement on planetary scales. This movement involved humankind coming to work on establishing and protecting a world that makes sense to people in their self-determination by becoming responsible together, in multiple forms, for the aggregate effects of planetary environmental change produced by the dominant political economies of the world. This movement involved making our institutions and cultures become accountable to whomever has been or may be dominated by the political histories and trajectories this moment of the planet has inherited, including colonialism, extant imperial forms, and the gross inequalities of actually existing capitalism. In other words, the movement looked forward and backward and involved many different workings of autonomy nonetheless aware of the moral need of a shared, planetary task. I called this movement after its core normative idea, "anthroponomy" – the collective self-determination of humankind as a whole throughout time.[1]

"Anthroponomy" was used briefly in the eighteenth century by Kant in his *Metaphysics of Morals* and is sometimes referred to as a rare anthropological term pointing to the ways in which human beings develop in their environment.[2] I know of no people who think of themselves as "anthroponomists." I can think of people who might be engaged, using other concepts, in anthroponomy. But even so, they are not united in that task.[3] The story of anthroponomy as a political movement awaits.[4] This book will try to envision what the strange word "anthroponomy" might mean in the life of a person initially disengaged from honestly confronting the persisting capitalist, still colonial, and often imperial world order.[5]

Now anthroponomy is helpful, because the concept of the Anthropocene erases geopolitical history and the specificity of the social processes that have constituted it. What must be made in *anthrōpos* (the human) is an opening for autonomous society of equal sense making, rather than the ongoing processes of domination and erasure in imperialism or the wanton inequality and wastefulness of capitalism, including their produced peripheries of exclusion and exploitation.

One of the areas of anti-imperial work is in addressing unfinished decolonization, and as aspect of that involves decolonial work. This is the work of cracking open the common sense of the world that supports ongoing colonization. "Colo-niality" is the name for this common sense, and the discourse of the "Anthropo-cene" supports it.

The "Anthropocene" is a form of "coloniality." The "Anthropocene" obscures domination, whereas moral accountability in the face of our planetary situation calls instead for autonomy. A consistent anti-imperial and anti-capitalist poli-tics, one committed to equality and autonomy, will engage decolonially with the "Anthropocene." Sometimes the way forward is the way backward. One name for this complex torque is "anthroponomy."

These last paragraphs contain many big words. But please bear with me for the duration of this book. I promise you that these words will become more life-sized.

~

As an act of imagination, this book isn't primarily a theoretical treatise. I will therefore beg your indulgence, as I have elsewhere,[6] intending that this book will make plain sense in the end. This book is an essay – alternately, a novel. Let me explain.

I mean "essay" in its original, sixteenth-century sense, and "novel" to pick out what the word literally means: something new. I want to first mark that the essay and the novel, as I will use them, have a relation to non-domination and to self-determination. That is why I am putting them forward here. When combined with – or seen ironically within – one of the major histories we have inherited, the essay and the novel have a relation to anti-imperialism – even to one of its spe-cies, decolonization.[7] They are a start in writing. They show a political possibility. They protest.[8]

The protest, here, is minor, beginning with one person. But in that person's life, the protest is major – a revolution in a way of life. It concerns how one will understand oneself and where one lives. It concerns how one understands one's political possibilities. And it involves how one should understand oneself in one's community.[9]

In this protest, the form of the essay matters as much as the essay's content. Essays write self-determination. Their point is wherever sense goes. What they produce is novel – the newness of sense as it appears in the life of the subject find-ing it. That finding has a name, "writing."

When writing finds a new political possibility that challenges a dominant social ordering and, being communication, shares it out in the open between people for consideration, writing is protest. It bears witness out in the open for all to con-sider.[10] What you have before you is a kind of testimony. It is a way of focusing my imagination around something that haunts me and makes my world almost pointless at times.

The form of writing called an "essay" gets its name from French *essayer*, to try. There is still an old verb form of this in English – to essay something. The Latin predating the French as its origin implied that one was weighing something. Trial, here, was evaluation.[11] But in the late sixteenth century, Michel de Montaigne wrote his *Essais* – the first so-called "essays" in literature – and they happened to be philosophy.[12] The sense of the verb began to move from evaluation to experimentation. Try something out.

Montaigne was a philosopher in the classical sense. Philosophy was a way of life.[13] It wasn't mainly theoretical. Montaigne was concerned that people question and find themselves, but his understanding of what it is to find ourselves was not ancient; it was modern.[14] Still, as the ancient philosophical schools did, Montaigne imagined practitioners – for him, readers – trying to use their lifetimes wisely. He imagined lives embodied, not bodies of knowledge disembodied. After all, although theoretical knowledge matters giving us a way to grasp our situation more fully, knowledge cannot properly be wisdom if it isn't integrated into one's own existence.[15] One needs know-how and a way of being with people to live freely, too.[16] Montaigne, aware of a European world increasingly deracinated and mixed through colonization and the early history of capitalism's transformation of most social relations, wrote to people in his class essaying a life that is hard to get right.[17] This was genteel, but it also opened up something commonly human.

The point is not to disregard contemporary academics, but to make sense of a way of writing that is close to living voices, everyday life, and to the strain and confusion of existing. It is to make sense of a way of writing where the process involves determining what one thinks and feels – thereby doing some work of determining oneself. When I read Frederick Douglass's *Narrative* or Virginia Woolf's *A Room of One's Own*, I hear philosophers more philosophical than many a professor writing analytical prose in a journal.[18] They write from their lives and they work out the conditions of autonomy – a life that doesn't dominate people and shut out everyday intelligence, a life finding reason and sense-making as something we all share. Their books are determined; they are even protests, and these uncommon philosophers write from wonder and self-determination with their whole being.

Learning from their spirit, I've written this book from a version of my life – refracted in the way Woolf's was, but not untrue to the basics. The essay quality shapes the inverted structure of the book. Its six questions end the chapters they title, and each chapter after them is accountable to them. For instance, chapter 1 doesn't answer the question, "How should I engage in community politics?" It *asks* that question, coming to it by the end. The next chapter then responds. But so does each chapter thereafter, focusing it through the twists and turns of the exercise. The last chapter bears the weight of all the previous chapters' questions. In this way, Chapter 6 can be said to answer Chapters 1 through 5, leaving only the question with which it ends. That last question is left to you.

My life shapes some of the essay too. I wrote from Cleveland, Ohio, once a land of many nations, a city in the state of Ohio in the United States of America, an imperial nation state. As a result, much of what I discuss in this book learns up from[19] indigenous scholars in the continent where I live, especially Native American and First Nations scholars and activists. That means there is an emphasis on their approach to decolonization and to decolonial work in the land where I live. I do not mean by this to de-emphasize the work of decolonization elsewhere. Rather, I mean to emphasize that the responsibilities I discuss are not abstract but demand relations where we live.

Similarly, when I refer to matters of my life that are personal, this is deliberate. It puts my relations forward. I deliberately did not document Native Americans or their practices using images. That isn't for me to do. I, rather, must come to terms with where I am and the specific location of the life I've inherited. From there, I must become accountable.

In other ways, however, this book is speculative. One can wonder about anthroponomy, and that wondering will consider something not yet seen, coming from the future. It is in this sense, mainly, that the book you have in your hands is a novel.[20] It seeks to produce a political possibility in a given life: the speculative fiction I call "anthroponomy."

~

Now the suggestion to link the largely *Roman* civic republican tradition of non-domination with anti-imperialism may seem odd at first. After all, it is from the Roman Empire that we have the word "colony."[21] But one of the ironies of cultures is that they can be at odds with themselves, even unbeknownst to themselves, like people. Then what a culture produces in one place may actually have implications that undermine or challenge what it produces in another, and, in the contradiction, a longing for a world that surpasses that culture may appear. Rome undermined itself when it developed civic republicanism. The question is how long it takes for this undermining to catch up with colonialism.

In *Between the World and Me* – a personal text and spiritual exercise–Ta-Nehisi Coates engages the "Dream" of "American" settler colonialism with its slaving and ongoing trajectory of racist oppression.[22] Near his conclusion, he links the history of colonialism with the path-dependencies creating runaway global warming:[23]

> Once, the Dream's parameters were caged by technology and by the limits of horsepower and wind. But the Dreamers improved themselves, and the damming of seas for voltage, the extraction of coal, the transmuting of oil into food, have enabled an expansion in plunder with no precedent. And this revolution has freed the Dreamers to plunder not just the bodies of humans but the body of the Earth itself. . . . [The Earth's] vengeance is not the fire in the cities but the fire in the sky. Something more fierce than Marcus Garvey is riding on the whirlwind. Something more awful than all our African ancestors is rising with the seas.[24]

Coates binds the capture and trade of slaves from Africa to the extraction of fossil fuels from the Earth – the first slaving colonialism, the second the core of settler colonial economy and infrastructure.[25] Redoubling the connection, Coates ends his diatribe by calling the automobile – used by whites to relocate to the suburbs and then commute, leaving Northern migration blacks stuck without equity in red-lined inner cities – "the noose around the neck of the earth."[26] In other words, an unsustainable, racist society based on path-dependencies to fossil fuels and structured by a history of domination is in the process of lynching the order of life currently on Earth.[27]

Calling on us to be speculative and angry, Coates's essay tries to puncture the "Dream" – to break through to a new possibility. He calls for something novel. The work of any such possibility must involve non-domination, throwing off the imperial patterns that have spread to the Earth itself and that imperil future generations along with the existing order of life on Earth. Non-domination, anti-imperialism, and climate justice intertwine.[28] Shall we go so far as to critique capitalism as well?

The civic republican tradition holds that people, morally equal, should not have to fear each other's arbitrary behavior or be subject to the senselessness of a social order. It demands that we should encounter human behavior, effects, and norms that are acceptable to us, that is, sensible by our own lights.[29] One name for this quality is *autonomy*.[30] Our social world should appear within norms we would accept, be thus acceptable, and in this way, legitimate and its various values legitimately authoritative.[31]

The civic republican tradition and autonomy go hand in hand. But our autonomy in our age of socially caused climate change is undermined by the path dependencies developed in capitalism and the colonial world system with its ongoing imperial struggles.[32] Future generations seem fated to be *pre*-dominated by our social order today and its basis in an aimless mix of capitalism, still imperial agendas, and the ruins and persistence of colonialism. By increasing the vulnerability of future generations, we expose them to the dominations of their day.[33]

Self-determination is critical in this conflictual scene. In self-determination, we seek autonomy. Having already challenged domination enough to determine ourselves, self-determination at the same time provides us with the sense to see and to challenge domination by developing autonomy. In the context of the sketch of history Coates gave, self-determination and anti-imperialism are, I think, modes of autonomy drawing on and promoting non-domination. What remains opaque is the extent to which Coates wishes to address capitalism as well as a mode of exchange.

Anthroponomy is a specific, coordinated use of autonomy to include humankind and its relations in moral accountability.[34] It is a mode of exchange based neither on profit, nor on power over others, nor on creating guilt and indebtedness.[35] Rather, its aim is to create a world with planetary scales of accountability.[36] People should be able to share a life that makes sense to each in their existences. This begins with not having their capacities to make sense out of existence dominated. In the context of the path dependencies that are with us as we enter the "Anthropocene,"[37] anthroponomy relies on non-domination and so on self-determination. It joins civic republicanism, anti-imperialism, and the critique of actually existing capitalism in a struggle that can appear in many modes and expressions of self-determination around the planet over time. This book's form is meant to *essay* one of those modes and to serve as one of those expressions.

Grounded as it is in autonomy, the story of anthroponomy cannot be written by one person. In fact, the book is dislodged into the stream of discourse through a response by Julia D. Gibson following the epilogue. This is only fitting for a book on anthroponomy. It doesn't get to remain closed. It has to be a story emerging across a collective involving disagreement. In this sense, the book in your hands is one perspective on a possibility, where what would truly be novel inside this perspective would be for others, using the irony latent in it, to disrupt this fiction with some sense of their own, some disagreement between worlds born of their self-determination. That planetary justice could be found in such good relationship is this book's actual, and not merely theoretical, proposition.

Notes

1 The first public mention of this idea of anthroponomy was in "Ethical Adaptation to Climate Change," *The European Financial Review*, October–November 2012, pp. 22–26, esp. the section "A strange new word." More recent public sketches of the idea include "The Fundamental Ethical Adaptation: Anthroponomy," and "Decolonialism and Democracy: On the Most Painful Challenges to Anthroponomy," *Inhabiting the Anthropocene*, June 8–July 27, 2016, accessed November 28, 2018. The first research publication specifically devoted to it was "'Goodness Itself Must Change' – Anthroponomy in an Age of Socially-Caused, Planetary, Environmental Change," *Bioethics (in Central Europe)*, v. 6, n. 3–4, 2016, pp. 187–202.

2 See David G. Sussman, *The Idea of Humanity: Anthropology and Anthroponomy in Kant's Ethics*, New York: Routledge, 2001 and "Anthroponomy," *The Free Dictionary*, Farlex, accessed November 29, 2018. However *Oxford Dictionaries* still locates the term exclusively in Kant's rare usage near the end of his life; "Anthroponomy," *Oxford Living Dictionaries, English*, accessed November 29, 2018.

3 People engaged in climate justice come foremost to mind, but so do those engaged with "Earth System Governance." See Frank Biermann, *Earth System Governance: World Politics in the Anthropocene*, Cambridge, MA: MIT Press, 2014; the conference program, videos and podcasts for *EJ20: Looking Back, Looking Forward*, University of Sydney, November 2017, accessed July 23, 2018; and also Stephen Gardiner's "A Call for a Global Constitutional Convention to Protect Future Generations," *Ethics and International Affairs*, v. 28, n. 3, 2014.

4 Cf. Allen Thompson and Jeremy Bendik-Keymer, eds., *Ethical Adaptation to Climate Change: Human Virtues of the Future*, Cambridge, MA: MIT Press, 2012. There, we call for the development of new concepts that are adapted to the moral and practical realities of socially caused, planetary-scaled environmental change. See also Geoff Mann and Joel Wainwright, *Climate Leviathan: A Political Theory for our Planetary Future*, Brooklyn, NY: Verso, 2018 – especially concerning the "adaptation of the political" and the notion (following Kōjin Karatani) of a new "form of exchange" which they call "Climate X."

5 The first principle of the platform of Extinction Rebellion is to tell the truth, where this is meant in the fully emotional sense of *admitting* the truth of the social world's dereliction to oneself and to others. See "Our Demands," *Extinction Rebellion* (UK), accessed September 16, 2019.

6 See for instance, *The Wind ~ An Unruly Living*, Brooklyn, NY: Punctum Books, 2018, also a critique of (Lockean) self-ownership, an issue that will haunt much of this book.

7 The history is European colonialism and then 'American' empire (i.e., the empire sought, maintained, and still projected into the future of the United States of America). Within it, one of the methodological assumptions of this book is that concepts from the *civic republican* tradition, such as non-domination, can be helpfully aligned with concepts from *anti-imperial* thought, such as self-determination. Cf. Philip Pettit's *Republicanism: A Theory of Freedom and Government*, New York: Oxford University Press, 1997; Glenn Coulthard's *Red Skin White Masks: Rejecting the Colonial Politics of Recognition*, Minneapolis: University of Minnesota Press, 2015. See also Eve Tuck

and K. Wayne Yang, "Decolonization Is Not a Metaphor," *Decolonization: Indigeneity, Education & Society*, v. I, n. 1, 2012, pp. 1–40.

8 Cf. my "Reconsidering the Aesthetics of Protest," *Hyperallergic*, July 12, 2016, accessed November 27, 2018.

9 This "one" is an abstraction that must hover behind the fiction of the book. I do not speak for everyone, not even for anyone else; I speak for my own life. Speaking by writing, I can begin to change myself and my relationships through understanding. This can suggest some of the things to consider as a human being.

10 Cf. Immanuel Kant, "What Is Enlightenment?" in James C. Schmidt, ed., *What Is Enlightenment? Eighteenth Century Answers and Twentieth Century Questions*, Berkeley: University of California Press, 1996, pp. 58–64, especially on the notion of the reading public; see also my "The Art of Protesting during Donald Trump's Presidency," *The Conversation*, January 20, 2017, accessed November 27, 2018. In using the term "protest," I hear Andrée Boisselle's work on testimony and indigenous ordering (law), where the testimony shared is important for what Kyle Powys Whyte calls "moral relationships," rather than being a "shout or an insult" as "protest" often appears in the public sphere. See Andrée Boisselle's Trudeau Foundation project issuing in her dissertation work, "Making Room for Indigenous Law in Canada: Toward a Reconception of Western Legal Theory," 2008, accessed November 27, 2018; Kyle Powys Whyte and Chis Cuomo, "Ethics of Caring in Environmental Ethics: Indigenous and Feminist Philosophies," in S. Gardiner and A. Thompson, eds., *The Oxford Handbook of Environmental Ethics*, New York: Oxford University Press, 2017, chapter 20; and my "The Neo-Liberal Radicals," *e-flux Conversations*, February 1, 2017, accessed November 27, 2018.

11 "Essay," *Oxford American Dictionary*, Apple Inc., 2005–2017.

12 Michel de Montaigne, *Essays*, translated by M.A. Screech, New York: Penguin Classics, 1993. "Happened to be" is subtext. Happenstance is a large part of Montaigne's philosophy.

13 Pierre Hadot, *What Is Ancient Philosophy?* translated by Michael Chase, Cambridge, MA: Harvard University Press, 2004.

14 Cf. Charles Larmore, "Michel de Montaigne," in M. Canto-Sperber, ed., *Dictionnaire d'éthique and de philosophie morale*, Paris: Presses Universitaires de France, 2004; cf. Charles Larmore, *The Practices of the Self*, translated by Sharon Bowman, Chicago: University of Chicago Press, 2010.

15 Cf. Pierre Hadot, "The Sage and the World," in Michael Chase, trans., *Philosophy as a Way of Life: Spiritual Exercises from Socrates to Foucault*, Malden, MA: Blackwell, 1995, pp. 251–263.

16 One could write "flourish" here instead of an italicized "live." But this would be to return to ancient *eudaimonia*, whereas Montaigne wrote in the stream of the invention of subjectivity where *freedom*, not expression of a culture's idealized meaning and sense of well-being, became paramount. Ancient *ēthos* was *eudaĩmonistic*; Montaigne's modern ethos explored autonomy and authenticity.

17 When philosophy emerged from the monastic communities in the Renaissance and early modern periods, it returned to antiquity through rediscovered ancient texts translated into Latin – for instance – from the Arabic translation of Greek originals in the eighth and ninth centuries, a.d. The idea of a philosopher as a person who searched for a wise way of life reemerged alongside, but this time through a growing sense of self-liberation in autonomy. Descartes was a soldier – but he was first of all a philosopher of subjectivity and did the meditations originally in between work as a mercenary. Spinoza was a lens-maker, but he was primarily a seer of intellectual liberation in discovering how one's life in the universe can make sense. Rousseau was a music-copyist; yet he orchestrated a world of consent and psychological independence, not domination. Montaigne was a small-time country gentleman. Note the conspicuous absence of professors. These figures each understood that to essay philosophy, whether in writing

or in mundane interaction, is to work to transform one's relations to the world and to oneself and, for any interested reader, to provide an opportunity for that reader to transform *them*selves and their relations to their world, at the least to consider what and how that could be on their own terms. See Hadot, 2004.

18 Frederick Douglass, *Narrative of the Life of Frederick Douglass*, Mineola, NY: Dover Thrift, 1995; Virginia Woolf, *A Room of One's Own*, New York: Mariner Books, 1989.

19 "Learning up" is an expression of Faith Spotted Eagle (Ihonktonwan Dakota). I discuss its meaning in Chapter 3.

20 I discuss in more technical terms the reasons for the personal *yet artificial* forms of writing in which this book's style has a home in "Species Extinction and the Vice of Thoughtlessness: The Importance of Spiritual Exercises for Learning Virtue," in Philip Cafaro and Ron Sandler, eds., *Virtue Ethics and the Environment*, Dordrecht: Springer, 2011, pp. 61–83.

21 "Colony," *Oxford American Dictionary*, 2005–2017. Joel Wainwright's criticisms about the relationship between empire and colonies in Rome led me to draw this conclusion about the apparent oddity of drawing on civic republicanism.

22 Ta-Nehisi Coates, *Between the World and Me*, New York: Spiegel & Grau, 2015.

23 On path dependencies, see John Dryzek and Jonathan Pickering, *The Politics of the Anthropocene*, New York: Oxford University Press, 2019 and on what runaway global warming might be, see, for instance, James Lovelock, *The Revenge of Gaia: Earth's Climate Crisis & the Fate of Humanity*, New York: Basic Books, 2007.

24 Coates, 2015, pp. 150–151.

25 Cf. Coulthard, 2015.

26 Coates, 2015, p. 151. Red-lining was a practice used by banks and realtors in the mid-twentieth century to indicate areas of "risk" where home loans should not be given. On a map, these areas demographically black, Jewish, and Italian were outlined in red. Cf. Tracy Jan, "Redlining Was Banned 50 Years Ago: It's Still Hurting Minorities Today," *The Washington Post*, March 28, 2018, accessed November 29, 2018.

27 The "order of life" does not mean all life, it means the current "rules of life" the more it edges toward a mass extinction cascade. See my and Chris Haufe's "Anthropogenic Mass Extinction: The Science, the Ethics, and the Civics," in Gardiner and Thompson, 2017, p. 35.

28 Such a breakthrough pulls rhetorically against the "eco-modernism" of the Breakthrough Institute, whose mission nowhere mentions environmental justice – or indeed *any* problems of oppression that we have historically inherited. See "Our Mission," *The Breakthrough Institute*, accessed November 29, 2018.

29 Pettit, 1997.

30 There are debates about whether republican freedom from domination implies autonomy or whether autonomy is necessary for it. As I will pursue further in Chapter 2, I believe that republican freedom does imply a relational form of autonomy and that such a form of autonomy is thus necessary for republican freedom.

31 Cf. Jean-Jacques Rousseau, *The Social Contract*, translated by Maurice Cranston, New York: Penguin Classics, 2003, especially its quest for a social authority that is legitimate.

32 Steven Vogel, "Alienation and the Commons," in Thompson and Bendik-Keymer, 2012, chapter 14; Mann and Wainwright, 2018. To speak of climate change as "socially caused" is to disambiguate our case from climate change that is caused by other than social processes as happened during the last ice age, and it is also to distinguish what we are seeing from something merely "anthropogenic," as if human beings as such have caused our calamitous situation today on the planet, rather than specific human societies and specific social processes such as industrialism and capitalism. I discuss this point further in Chapter 1.

33 See my "Presentism the Magnifier," IAEP/ISEE, University of East Anglia, June 12–14, 2013.

34 I will discuss this technical matter by Chapters 5–6.

35 Cf. Kōjin Karatani, *The Structure of World History: From Modes of Production to Modes of Exchange*, translated by Michael K. Bourdaghs, Durham, NC: Duke University Press, 2014.
36 For instance, of space and of time. This technical matter will be discussed beginning in Chapter 1.
37 Cf. Dryzek and Pickering, 2019. We should be careful here to include the history of colonialism that troubles the very concept of the "Anthropocene," something I discuss in Chapter 1.

Figure 1.1 The previous owners chose wallpaper for the German house, Shaker Heights, Ohio, 2019

1   How should I engage in community politics?

Eight P.M., Sunday, home, Shaker Heights, Ohio

How troubled my mind has been recently, even last night as I slept. I dreamed fit-fully. Like a drip from a faucet in the night, the doomsayers unsettled me, telling me that everything human is up in the air; civilization as we know it is finished.[1] But the problem is: I do not think that this global "civilization" *is* civilized.

I don't know if our globally interconnected order has even the requisites of a civilization. Actually existing global capitalism isn't a society; it feeds on culture, imitating one; and to call it "advanced" is offensive to morality.[2] Global capitalism's "planetary" flows are even less civilized, no matter how socially organized their unintentional and negligent production.[3] Can we speak of "civilization" when we face an impersonal, acquisitive inertia?

Moreover, people have been living *under* the once-and-still colonial world system's hegemony for centuries. The shape of the international order bears the structure of European colonialism. Self-determination struggles arise in the terms of the nation state, the political form that structures international order, rather than in the terms of indigenous political organization. And there are many still-colonized nations where indigenous communities live under colonial states. My nation state is one of these colonizers.[4]

Looking at my own nation, the United States of America, I see the extraction of resources from lands and the capture of people for labor trailing back to when the first colonial conquests began. Some of my nation's most eloquent writers have spoken out at length about the enduring effects and persistent patterns of this colonialism.[5] It is hard to fear the loss of a "civilization" that has never been civilized.[6]

Still, the doomsayers are voicing a truth that needs to be heard: our planetary environmental situation is precarious.[7] Global political economic disorder appears insufficient to address it.[8] Yet not even social science can predict a truly political event, a revolution, say, or a collapse. It can explain much in retrospect, and it can help us conceptualize the paths actually existing societies seem likely to take.[9] One has to be careful with sooth-saying – not because it is depressing, but because it is *metaphysical* in the objectionable sense Immanuel Kant took on in the *Critique of Pure Reason* at the dawn of the critical age of European modernity. Sooth-saying slips into knowing beyond its bounds. It stands on the one side of

Kant's Third Antinomy of freedom verses determinism, claiming that society is determined.[10] But this is to overlook the power of reflective consciousness and organization to open up possibilities in society for living. We shouldn't assume that we are blithely free – that would be objectionably metaphysical too – but neither should we write as if humans from this point onward are unable to change the world. The question is how to relate to the truth of our precarity without becoming fatalistic or all-knowing.

That being so, I think it's actually immoral to act as if we can't change this world, at least if intrinsically fallacious determinations lead us to think that our decisions don't matter, here and now. When it comes to my own life, I don't have the moral opportunity to focus on doom, much less the loss of a wicked civilization. I have a different problem: apathy caused by not knowing how to be, how to *respond* to the situation my wicked civilization is in. This is not a problem of nihilism, of meaninglessness. It is a problem of disorientation. I'm looking for how we ought to change the world and for the direction in which to head. The grief I may feel for the losses we are likely to face and for the load of delusion we've been fed by a wicked order is real, but it must become part of living. I have moral relations I cannot let go.

I want to know where we should be aiming as people. I want an idea to organize my action and shed light on what should be, not what is. At the beginning of the day and at its end, I have to be accountable to myself – to what I know is right. I can't just sit by knowing that I should do something but not having a clear idea of what it is. I know that I am responsible to make things better, not sit idly by – or sink into resignation.[11]

There is a kind of apathy created not by loss of heart but by conflict and confusion.[12] I know that we who live in this actually existing global economy and its colonial state systems, especially all who will be affected by it, are morally required to construct a different world. I am apathetic, because it is hard to see clearly how I should think about that world, what ideas I should find morally acceptable to animate my pursuit of it. The apathy interferes with my accountability, as if I were a fibrillating heart.

When people watched television through antennas on the TV set, sometimes two channels would cross over each other, creating a screen split by competing images, disrupted and starred with static. My apathy is like that.

The problem I'm facing is a specific kind of utopian problem. To say that something is "utopian" is to say, literally, that it exists "without a place" – nowhere. The different world that I should be working toward is utopian not simply because it does not exist yet, but because I do not have an idea of what kind of world it should be. Perhaps others do, and I must seek them out. But tonight, I want to see what I can do to clear my mind myself.

I find myself acting on the edge of pointlessness. There is something that has to be done. Not to try to figure it out makes it hard for me to face myself. One point in struggling is to have some integrity.[13] The point is to try to use my life for some good, even if I cannot see my purpose clearly. To not try would be to be a bad member of my generation.[14] Even more minimally, the point is to help show others, better positioned than me up ahead, how to try – or how not to! The point is to spur some imagination of a different world. The point is all of these things and more – for instance, to produce *some* change during my lifetime, even if it isn't sufficient. That would be a partial success and a partial failure. Still, I do not have a clear idea of the different world we should construct. Maybe others do.[15]

Being alone in this fibrillating state makes me angry. Anxiety reveals the possibility that the norms of the world are up in the air and suddenly appear contingent. Some even call this anxiety an "experience of freedom."[16] I think of it as a dim awareness of our power to construct what makes sense in our own terms. My anxiety concerns a political problem. We do not have to be here in this so-called "civilization," but we are here because of norms upheld and decisions made across societies. It's facile to say that we could simply choose a different world as if consent is all that is the issue, but we enact the world that we think makes sense and we are responsible for asking whether our world does make sense. We are accountable to the realization that it might not. This anxiety is underneath everything, the ever-present possibility that our society could be otherwise and is contingent.[17]

Many of the processes in my society – and in societies bound by shared systems around the world – are shifting the organization of the biosphere – altering the rules of life – and will change the geology of our planet.[18] These changes include global warming, the risk of a mass extinction cascade, and the pervasiveness of industrially produced toxins throughout the biosphere. They include the expansion of industrial animal farms for meat and dairy and the accumulation of plastics within soils and waters on a scale exceeding the imagination. They include many more changes, too. We are said to be entering the "Anthropocene."[19]

The "Anthropocene" is a marker for a bundle of biochemical, biological and imminent geological effects, systematically interconnected, which appear to be an unforeseen consequence of human civilization as such. So the story goes, humankind has caused the Earth to reorganize as if a trauma occurred to a system, as if we were a gigantic meteor, oxygen in an anaerobic planet, or some such geological force.[20] Now we are responsible for something for which we didn't know we could be responsible.[21] Now we change the rules of life on Earth despite our intentions, alter the whole planet without purpose. What is worse, our record says that there will be little but chaos in the end.[22]

But *our* record? I have doubts about the word being hurled into our life – the "Anthropocene." It speaks for all of humankind. According to it, humans – *anthrōpos*, human being in Greek – are the cause of the effect that the name marks.[23] Clearly, though, that is an obfuscation of the truth, and those of us who want to tell the truth shouldn't be saying it.

Only *some* societies – and in particular, some specific systems – have generated the big problems that are most commonly discussed in explaining the "Anthropocene."[24] *A* world caused this, not *the* world. Industrial energy driven wantonly on by capitalism has largely driven global warming.[25] Large scale monocultural agriculture, deforestation, factory farming of meat animals, overhunting and overfishing, the prevalence of industrial toxins throughout the biosphere, and habitat destruction caused by expanding urbanization especially in capitalism are the main drivers of the elevated extinction rates that risk a cascade into a sixth mass extinction since life began on Earth.[26] Industrial society looms large here, implicated as it is in capitalism and its history and persistence of extractive, settler colonialism.[27] The point is, humankind as such is not the cause of some of the main problems of the "Anthropocene."

So why press the word? The shadow of corruption appears: The "Anthropocene" can understandably be experienced as colonialism's shadow, extending far into a new geological epoch.[28] Colonialism already did on a smaller scale what the "Anthropocene" is doing: devastating ecologies, forcing migrations, producing extinctions, polluting the waters, introducing invasive species, and leaving the vulnerable – those who are not in power – with unacceptable conditions stretching into the future, burdening and harming their future generations.[29] Moreover, there are serious arguments to the effect that the "Anthropocene" ought to begin with the rise of the colonial world-system during the "Age of Exploration" and the advent of mass, European colonization of much of Earth.[30] After all, it was those social processes that extracted vast amounts of natural resources and slave labor for the accumulation of great wealth in the coffers, banks, and properties of the already wealthy nobility and nascent bourgeoisie in societies that would then be able – based on those resources, wealth, and scale of power – to industrialize.[31] It was those social processes that started the world system violently being abutted – shearing, smashing, shuffling, reorganizing ecology and polity to favor the ones in power against the vulnerable wherever they lay. The edges of monarchies became the bounds of national territories carving up and placing in tension the lands of the Earth as the colonizers' property.[32] These social processes opened the way for global capitalism on the one hand and a state system made of once and still persisting empires on the other.

The word, "Anthropocene," is at the very least bad recording; it is imprecise.[33] But it is avoidance, too, since it obfuscates morally important considerations, such as the source of the problems we face. Why should we trust a name that spreads blame away from obvious causes in the midst of systems that avoid accountability even to the point of being arbitrary?[34]

Wednesday evening, home, Shaker Heights, Ohio

The name, "Anthropocene," is a barrier. It skews the sense of our situation. The planetary situation in which we all live is worsened by the sense being given to it, and the sense being given to it – the "Anthropocene" – expresses unintentionally some of the very dynamics that created the situation. The shifting of this planet's biochemistry, biology, and geology as a result of industrial processes at a global scale over hundreds of years, compounding even more centuries of violent habitat destruction and reconstruction in colonial exploration, itself settled into thousands of years of monocultural practices at non-industrial scales, and a millennium or so more of killing megafauna to extinction . . . all that is *worsened* by implying that humankind as such is planet-altering. A mere piece of semantics worsens such massive processes, because it obscures causality, and that obscures *accountability*, retrospectively across history and prospectively toward social and political action that would be sufficient to make our situation live up to our commitments.

Worse still, calling our situation the "Anthropocene" and meaning thereby that humankind as such is a geological change agent[35] belies what is called *reification*, something that has become a feature of societies structured by industrialization or capitalism.[36] Reification – from the Latin *res* (thing) – takes "a relation between people" in a social process and turns it into a "thing."[37] This then obscures both the social process and the potential for autonomous relationships between people whose consent or coercion supports the process's continuance. Specific social processes to which people have consented or into which people have been coerced – and dominated – are the causes of our planetary situation, *not* merely fixed "laws of nature." The processes in question include colonial world systems, industrial economies, neoliberal risk-taking, and more. The cause of our situation is thus not simply being human, that biological thing. Calling our situation the "Anthropocene" and thereby implying the cause of our situation is humankind as such serves to obfuscate the specific cultures, institutions, and practices of people, including the reasoning behind the social choices and social abuse that have brought us, often in abjection or loss, to where we are now. It also obscures our autonomous potential going forward from here on out to be responsible for our social processes.[38] Most disturbingly, it represses the people affected by those choices and that abuse who would *reject* them. By reifying the social relations causing our situation, the "Anthropocene" silences both history and people.[39]

The problem goes deeper, too, because the name, "Anthropocene," builds apathy into the situation by making being human into a force. Karl Marx called this "alienation," specifically of the kind that takes specific social relations – he was thinking of capitalism – and makes them into human nature.[40] But it is also what Jean-Paul Sartre called "bad faith," for as human beings we are structured by relations of authority, not simply of force.[41] When free, we reason, consider things and act on what makes sense to us. When not dominated or coerced, we are autonomous beings in spirit.[42] We form intentions, aren't simply pushed about – or so we think in order to have a recognizably moral life. Moreover, when we discover that we have been pushed about by others, we take that as an indignity and wrong to us.[43] The name, "Anthropocene," however, by alienating our autonomy into a claim about humankind producing our situation makes it seem unnatural that we could choose to change things collectively, that we could become collectively autonomous and morally accountable for our situation. In effect, the "Anthropocene" subtly eliminates the moral point of view.[44]

For any society committed to scientific accuracy, reification is fallacious. But when it serves to deflect attention away from the social and political order that benefits from maintaining its production of the situation we face, passing on reification is complicit in corruption.[45] What kind of corruption? Thinking of it, I would call it at least as a "social process obfuscation."[46]

What could that be? A social process obfuscation is any sense given to things that covers over, dilutes, minimizes, or misdirects the extent to which – or the form in which – those very things are the result of a social process open to social change. Reification is kind of social process obfuscation. But so is making all of humankind responsible for what imperialism, colonialism, capitalism, and industrialization mainly produced.[47] The "Anthropocene" obfuscates these social processes and as such appears complicit in their wantonness and injustice.

Thursday evening, home, Shaker Heights, Ohio

I worry that there are corrupt incentives for accepting the claims of the "Anthropocene." The world system's economies are fossil fuel based; their form is industrial still, networked by the Information Age's interlinking of production, distribution, and waste disposal around the globe with slight delay.[48] Global capitalism works increasingly by speculative investment in futures that gamble with risk. Its logic – called "neo-liberal" – avoids accountability and privileges opportunistic and short-term strategies.[49] All this is permitted by an international political order that is still stamped colonially in its denial of indigenous nations, repression of colonial history, maintenance of inequality between colonizing and colonized nations,[50] and which is resistant to international accountability, especially regarding future generations.[51] This order is still an order of intended and persisting empires. It is convenient for vicious social systems to spread blame *away* from them to the point of obfuscating the responsibility they have for the continuing dynamic of the world's disorder. The "Anthropocene" as a linguistic act seems to enable imperialism, colonialism, capitalism, and industrialism to continue when we think about the environmental situation of our planet.

When I think about it, the contradictions here are startling. Many of the things the world systems involving my society are doing to the planetary environment appear unjustified from the standpoint of some of the main express norms of my society, even of the major moral agreements of parts of the world systems. Yet the unjustified inertia continues without explanation, let alone justification.[52] When, for instance, we realize that we are putting at serious risk future generations, we contradict the commitment to uphold human rights. When we see that the world of life is being treated wantonly with destruction occurring unintentionally, massively, and without thought, we contradict the widespread commitment of the major religions of the world to be thoughtful with life. When, in addition, we see that the major agencies of the world system erase the meaning of the land for many indigenous societies around the world, thereby erasing their moral grounding and sense of dignity in the relationship they enjoy with the land, we realize that even liberal norms of respect for people's ways of life are being contradicted – let alone norms that are indigenous and deserve widespread support insofar as they are non-dominating.[53] Thus, what is systematically allowed lacks the integrity of many people's convictions – and of many societies, cultures, religions, and ways of being in the world too. If the systems were to become morally accountable, they would cease business as usual and would have to reorganize.

But when I look at the nation states permitting their contribution to and influence regarding the planetary, environmental shifts underway, I see that even when they address global warming – the major event of the "Anthropocene" so far – they act insufficiently and seem to have difficulty including civil society in their deliberations, especially including those people, groups, and representatives of societies who are upset about historical and looming injustice, including the history of colonialism.[54] This in and of itself is a problem of accountability both concerning results and to people with a legitimate concern. It leads me to question the extent to which such nation states have an incentive to avoid moral accountability. I then wonder what their relation is to the dominant systems responsible for our planetary situation – imperialism, ongoing colonialism, capitalism, and industrialization.[55]

The point is, the "Anthropocene" – despite being inaccurate – is too close to systemic moral ambivalence for my comfort. If the sense being ascribed to the situation we are in contributes to the problem of that situation, shouldn't we find a different sense – a different understanding of the situation – to help us deal with it? If the problem is a planetary environment our world system has unintentionally caused that fails to cohere with moral convictions basic even to that world system, the major religions of humankind, and the cultures of many indigenous nations, shouldn't we understand that problem in such a way that our power to cause things could cohere with basic moral convictions across the world system, the major religions of the world, and the cultures of indigenous nations? Moreover, if the world's systems have unintentionally caused a dangerous and unjust situation by perpetuating injustice inherited from the imperial and colonial world system, short-term thinking and a selfish drive inherited from capitalism, and dangerous methods of energy use, production, and waste inherited from industrialism, shouldn't we articulate our capacity to restore moral relations in the world, specifically relations of justice? After all, the only thing that would appear to be acceptable to and for beings who are autonomous in spirit would be systems that do not render us heteronomous, that don't dominate any of us, systems where participating in social processes won't contradict our convictions and perpetuate injustice.

What more can we say about such systems? What would characterize them? And what would be the social processes that would allow us to articulate and make them? They would certainly be different from what appears to be the truth now: vicious systems that push aside and disregard people's moral convictions for the sake of selfish ends.

Friday afternoon, office, Cleveland, Ohio

Social process obfuscations are different from ignorant inaccuracies. They involve sense-making where we ought to know better. So they lead one to look for systems of domination and privilege to which people adapt, shaping their sense of things so that they do not disturb a charged and dangerous social reality. When things are off limits in people's minds even though they are morally demanded on reflection, the deformations of domination are often present. People shape their sense of the world around what appears realistic to them,[56] and challenging massive systems of domination or privilege either which may threaten them or from which they benefit hardly appears realistic.[57] In the case of the "Anthropocene," the incentive to deflect away from systemic failures of moral accountability by making all of humankind a reified causal agent isn't hard to find, nor are the systems of domination and privilege.

Because of the extent to which the "Anthropocene" involves us in heteronomous systems, being accountable for the "Anthropocene" doesn't appear mainly to be a matter of being accurate and of having integrity. Systemic autonomy seems more basic and important first. To say that our problem is mainly a lack of virtue as some have implied is not theoretical or political enough.[58] It doesn't help us understand our particularly twisted situation on Earth now.[59] The planetary situation we are in is novel. However, it isn't caused simply by "humankind." Moreover, any sense we give to it should reflect our capacity for responsibility, not redouble our reification.

If autonomy is basic to being human and if domination is a wrong to us as autonomous beings, then we would want a concept making demands on our situation that involves the systematic autonomy of humans.[60] We would want this concept to help people collectively determine the sense of the situation in which we find ourselves. I propose that this concept, and the process to which it refers, is called *anthroponomy*.[61] Anthroponomy is a social process wherein the effects of our world systems become justifiable to us across humankind.[62] It is a name for a process that restores integrity and accuracy.[63] But anthroponomy isn't simply a virtue. It is a systemic condition found across social processes by way of diverse practices, institutions, and cultural beliefs.

Anthroponomy is a social and historical process. It is the opposite of what the reification involved in the "Anthropocene" suggests. Moreover, opposition seems right. Suppose that instead of understanding our planetary situation as evidence that humankind as such has become a geological force, we took it instead to challenge us from here on forward, to change our social processes to become anthroponomous? What would such systemic autonomy involve?

Sunday afternoon, home, Shaker Heights, Ohio

On the face of it, anthroponomy seems to call for prospective, political responsibility for our shared planetary situation.[64] The main responsibility involved in it would seem to be to collectively make social systems that are justifiable to the people affected by them.[65] This would involve not dominating people and social, political, and economic forms of power that are legitimate to those affected by them.[66] Given that our planetary situation now involves multiple histories of violence and domination, anthroponomy would also seem to include supporting, restoring, and protecting the autonomy of people against the continuing dynamics of these histories and their ongoing systems. Then not only would world systems that perpetuate injustice be deconstructed, but the social processes we adopt would support the restoration of the autonomy of those societies especially that bear traumatic histories of domination and coercion.[67]

There are so many things to be worked out about such an idea, however. How can anthroponomy fit something as vast as humankind and yet show up in any given system? It is hard enough to find ways to develop shared autonomy in a family, let alone an organization. It is even harder in a city, harder still in a state. In the existing world-system, developing collective autonomy in international matters seems next to impossible to realize, as global warming conventions have shown.[68] What would it mean, beyond that, to develop autonomy across humankind in such a way that humankind had in view its anthroponomy? This problem leads to a second.

When people work out their autonomy together, much depends on their being able to address each other. Accountability is bound up with autonomy, and contemporaneity seems bound up with accountability in situations where shared autonomy must be worked out. How, then, is anthroponomy possible with respect to future generations, those with whom we cannot meet to work things out? Yet the planetary situation in which we find ourselves puts future generations at severe risk.[69]

When it comes to future generations, the most obvious problem seems to rest on the nature of time. In this way, it appears to be absolute. Since when is it possible to act effectively in ways that contravene reality? Take a morally relevant definition of a generation: it would be one where the new generation has no actual contact with your own.[70] Imagine, for instance, that you are speaking of the morally relevant generation as being the one that includes your great-great-grandchildren, were you to have that lineage. In a world where life-spans increase, imagine that the generation includes your great-great-great grandchildren, and so on. The point is, it is strictly impossible for both generations to conduct reciprocal relationships. This is due to the "causal asymmetry" of the situation.[71] Your generation can affect the future one, but not vice-versa. Reciprocity is impossible, and you can affect their fate, but not vice-versa. This relation in time is prime for injustice. It is easy for *us-now* to fail to mind *them-in-the-future*, to pass problems along to them or to burden them with the costs of our lives. Think of global warming, where we offload the costs of a carbon-intensive, industrial, globalized economy on future generations.[72]

Our current generation *predominates* over future ones. People in the present have power to drastically affect people in the future without it being possible for those in the future to hold those in the present to account. Even if we develop institutions that allow us to delegate individual responsibility to transgenerational, collective orders open to change,[73] still the nature of time forecloses our working out legitimacy across generations together, although that might appear morally needed for anthroponomy. Moreover, in predominating over future generations, we can make them more vulnerable to the dominations of their day.[74]

What would happen at best, it seems, would be for us now to make decisions that those in the future should find legitimate. But given that involving anthroponomy in our systems now appears to come up against histories of violence that are ongoing, the suspicion of speaking for others in the future and repeating dominating norms is understandable. A feature of the imperial, capitalist, and industrial world systems has been the development of technologies, including specialized discourses, for managing populations – slaves, the colonized, prisoners, laborers, those with abilities as opposed to those with disabilities, even those who became coded as "abnormal."[75] These technologies have become so bound up with the process of determining norms in society that core features of dominating world systems' ideas of what is normal are expressed in them. Then the "person," the "individual," the "self," the "autonomous agent," even the "human being" all became coded ways in which dominating systems manage their domination of the colonized, exploitation of the laboring poor, tacit control of those benefiting from the systems, and marginalization of people who cause difficulties for the smooth operation of the systems, either due to their unproductivity in a given system or their way of introducing space into norms at basic points – such as gender – so that the social world begins to appear as arbitrary as it often is. Oppressive normalization becomes implicit in everyday nouns we use to name ourselves.

On this way of thinking critically, our oppressive, normalizing systems create the concept of humankind as if it were a generally visible reality. In this way, they can manage the norms of being human, including determining, suggesting, or excluding those who are non- or sub-human.[76] For such systems, anthroponomy might seem a pretext for further consolidating "humankind" as a normative order that excludes those who are not human or less than human[77] and which disciplines people into a proper humanity.[78]

Under this suspicious lamp, it becomes doubly concerning that anthroponomy involves *nomos* – the creation of norms, customs, laws, and so on. Anthroponomy's abstractness becomes sinister, because it seems to allow the universalizing machines of world systems to construct humankind's proper functioning without input from those whom they exclude.[79] Could "anthroponomy" be a pretext for a disciplinary and disciplining construction of "humankind" as a stable entity, although people are always more than the "policed" order of such a stable set of identities?[80]

Sunday evening, coffeehouse, Cleveland Heights, Ohio

Now one thing clear to me is that any workable concept of autonomy between people should be constructed around the possibility of disagreement. The idea of autonomy between people would seem, at least in the abstract, to begin with the possibility of disagreement, not agreement. For people to find anything legitimate, it must first be possible for them *not* to find something legitimate. One way to put this point is to say that the right of refusal precedes the assumption of unanimity. To talk about two entities collectively sharing norms, then, is to say that they came to share them, across a process that began with the right of refusal.

Let us say that whatever anthroponomy is, it is built on the priority of disagreement. This positions anthroponomy not as a top-down, administrative philosophy whose sense of norms involves making people normal and manageable. Rather, anthroponomy becomes disputatious, bottom-up, often strange and ultimately collaborative. Anthroponomy, rather than being on the side of the "police," appears to be on the side of "the part who have no part" – those who have been excluded from a society and organize as a result of suffering injustice.[81] Anthroponomy appears to be on the side of the colonized and those suffering from the legacy of imperialism – as well on those who are marginalized. Insofar as people might make claims on behalf of lands or more than human beings with or toward which they live out moral relations, anthroponomy appears on the side of the irruptive disagreement with, for instance, capitalist and settler colonial property relations excluding the moral being of lands or with industrial production dominating more than human animals for the resources that they provide meat eaters, pharmaceutical companies, and science.[82]

What of "humankind," then? Obviously, "normalization" is sophisticated when it comes to how people are determined to be who they are, what they think makes sense, and where and how – with what practices, for instance – they think that they can share things. But the general problem in the critical suspicion seems to be that humankind is constructed *against* or *prior* to disagreement.[83] The assumption I've just articulated about anthroponomy changes that. If the possibility of disagreement is primary to anthroponomy, if disagreement structures anthroponomy as a process, then whatever humankind is, it emerges *only through* and *not despite* disagreement and its persistent possibility. We might go so far as to even say that whatever "humankind" is, it is they who appear in the field of norms *as* disagreeing and those *with* whom they disagree. In this way, humankind is constitutionally unsettled. Humankind is disruptive and related – until the next unsettling. Anthroponomy is a philosophy of unsettling, not of settling.

Here, then, we come straight again to the problem of future generations. If there is a "part who have no part" in our present society, it is surely generations far into the future. They cannot have a part, not simply because they are not here yet, but because to assume that they will agree with our order of things flies in the face of history, contingency, and disagreement. People have to decide to do, in time, what they have and want to do with their moment. Time builds the possibility of disagreement into society. With aging and death, new people are left to decide. The problem of predomination is also the reality of disagreement in time. Time obliterates normality. It erodes normalization as the sand covered Ozymandias.[84] In time, new exclusions appear as people shift their sense of the world. These must then be challenged for "humankind" to emerge again from underneath hegemony. Disagreement is a feature of time.

We have an experience of predomination in every life. It's that of being predominated over by past generations. Consider, first, a negligent generation that messed us up in carelessness and selfishness. Then consider a generation that, yes, didn't know what we'd face in the future, was often ignorant, and had hard to reach biases ingrained in their society. Yet it tried hard to respect us in our future, even to make our lives better off. The question isn't whether they were perfect – eternally and absolutely – but whether they lived, passed things on, and set things up for those who came later in a considerate and thoughtful way, acting as morally as they could toward the future. Just as the question is whether the past displays a moral relation with us today, so the question is whether we can display a moral relation toward the generations which come after us far into the time's reach.

Since anthroponomy begins with acknowledging disagreement as basic – since it presumes the possibility of disagreement as an ever-irruptive event challenging "us" to confront "them" suffering injustice by "us" – so we should presume that future generations will, in some unknown way, disagree with us. The question is what form our accountability now around this presumption should take. Perhaps the institutions we need are particular kinds of moral institutions, ones that do not presume to decide for the future but which hold us to integrity and thoughtfulness? Perhaps they must keep themselves open to disagreement as thoughtfully as we can make them?

To help answer these questions, it seems that I need to develop a philosophy of disagreement that can help me appreciate disagreement around the planet in different ways and across time in different histories. It seems, too, that I should look for moral relations in disagreement. If only in disagreement is the possibility of legitimacy, autonomous practices and institutions should be shaped accordingly – not as policed spaces of determination, but as spaces of accountability that evolve during times of disagreement. Anthroponomy would become a philosophy of accountability in disagreement built around challenges to exclusion, oppression, and domination. Just so, it would be constructed as changing in time for reasons of justice, moral rectification, and moral restoration. If anthroponomy is a moral philosophy of disagreement, then the things to stick with in order to see anthroponomy clearly are the actualities and manners of autonomous disagreement across and evolving within humankind, itself ever open to being shaken up by those who have been excluded. As I consider anthroponomy more closely, I want to try to discern some of these open forms, beginning by understanding disagreement's relation to autonomy more closely.

A brief moment on Tuesday, a land along the lake, once shared by many nations

Why would we think that anthroponomy is one thing or one discrete process? Perhaps it is many processes, just as there are many ways in which disagreement can appear with those who have been excluded or oppressed. Perhaps anthroponomy is fractal. There could then be a bit of anthroponomy in a family just as there could be a scene of anthroponomy at stake in a state, between states, or against the state.[85] Anthroponomy would appear in a specific kind of accountability with disagreement, not in unanimity, and there would be many ways in which such accountability could appear in the many areas of life where some of us live silenced, hidden, downtrodden, or removed.[86] Anthroponomy would then be many processes joined by a family resemblance[87] with one thing in common: the orientation to create social processes accountable to exclusions, oppressions, and domination across humankind so that a planetary situation involving and producing injustice can be addressed, rectified, and improved by seeking legitimation in the eyes of those affected by it. Anything less would leave some at the mercy of others as all of our environments change due to the social processes of only some of the world's systems.

Wednesday early morning, Shaker Heights, once land of many nations

I like my general direction. For the second time in a rough week, it's slight progress. But to talk of anthroponomy being fractal is one thing and to understand it as practicable and plain in my community is another. I began this essay wanting to dispel apathy constructed by confusion in my society. I've often been inactive because my purpose isn't clear, not because I lack energy, desire, or moral motivation to act. To dispel my apathy, it won't be enough to keep anthroponomy up in the sky in abstract remoteness. Anthroponomy must come down to earth where we all live. I need an approach to essay it in my community, even if what I am trying is a fractal form and a general orientation to social construction. I need the forms and the constructions to appear at the pace of living.

What is my way into community politics? My reflections over the past ten days led me to wonder whether anthroponomy is a bundle of processes sharing a family resemblance. I drew an analogy to a fractal form. Thinking of anthroponomy as a process that could appear in small or at vast scales makes me wonder how it might be different than autonomy except in its clear goal of including humankind in some way. After all, my reflections switched back and forth between autonomy and anthroponomy as if the two share a form, but autonomy, unlike anthroponomy, is something life-sized, an everyday word and concept. Perhaps autonomy can point me on my way into anthroponomy. Autonomy is a word that makes sense within community discussions, including political ones. For all that, too, it isn't a word that seems to stifle disagreement.

Yet "autonomy," according to my dictionary, is misleading, given how I've learned it. The *Oxford American Dictionary* says that the word applied to a *society* means being self-governing, whereas the word applied to a *person* means not being influenced externally.[88] These are actually different ideas. A country can govern itself even if it is influenced by, for instance, weather patterns. Governance responds to such influences. Similarly, a person can govern their own lives without them being entirely unaffected by what is around them. Why else would they govern if not that they had to respond to what affects them?[89]

Convention is confusing, but the root of the word "autonomy" isn't: it means "self-law" (*autos* – self – plus *nomos* – law – from the Greek). The idea behind autonomy, I think, is that what governs you is something you should be able to endorse. What you feel compelled to follow should be what you think has legitimate authority – it should make sense to you. You should hold, in effect, a statement such as this, "I believe, not begrudge, that this is law" (or norm). In other words, when I am autonomous, I think that the norms I hold are actually normative – right, good, true, or some other value of legitimate authority to me. They are truly authoritative to me.

Also, when I am autonomous, I'm not simply conforming my sense of life to a constrained existence in society that I wouldn't otherwise choose if society were constructed differently. That would be an in-grown form of hctcronomy – settling for what doesn't make sense although I have the space to freely examine and explore it and other options.[90] When I'm autonomous, the things that guide my life cohere with my sense of life. I can hold my head up, have a clear mind, try things out, explore enough of the world to not simply settle for it, and judge life without intimidation. My life's guidelines are truly justifiable to me. They make sense, and I believe that, were others to see my perspective, what I think makes sense could make sense to them. I find it reasonable, which means that I think others would too.

So initially, let's say that instead of what is in the *Oxford American Dictionary*, autonomy is the condition – and power – of living with the opportunity to think for yourself, where your living truly makes sense to you in a way that could be justified not only to yourself but also to others. Otherwise, it wouldn't be justifiable. Autonomy doesn't exclude others in its sense. It is inclusive.

Being so, there is a sense in which autonomy must then be open to justification given anyone's disagreement, including that of, or on behalf of, any morally considerable being.[91] Since one's autonomy must always then be potentially accountable – open to one giving an account of what one thinks makes legitimate sense – autonomy is intrinsically and potentially public. It is, we might say, intrinsically disagreeable – about to be disagreed with. It is out there in the open, plain and available, not shying away from disagreement and accountability.

This autonomy makes sense to me. Autonomy is structured by and displays the kind of accountability we should expect, we who live our own lives and are morally considerable. We should expect a society and communities where people and institutions are accountable, available to our claims for justification should they appear objectionable, and where the idea is to make sense to and with a community, oneself and others included especially in and through disagreement. Community politics isn't far off here.

Given what autonomy is, then, disagreement – including disagreement in our community – appears to be the moment where autonomy comes into its own. There, outrageous inconsistencies between our commitments and our institutional and social effects can't be avoided.[92] There, things that we are doing that don't make good sense before others can't be things that we simply ignore before others, stonewalling them. Autonomy shouldn't be a castle; it should be open, all around, since it is the process – the condition and the power – of a life that truly makes sense to oneself and before others.[93]

To be practicable and plain, I'm going to follow autonomy and see how it can take me to anthroponomy. From there, I'm going to see how anthroponomy can take me to reconceptualizing the situation in which we find ourselves on Earth. In that way, I hope to end up with a way of living in my community facing the planetary situation in which we are deposited as a result of the historical crime and senselessness of the world's systems in many respects.

My goal, for now, is to engage in community politics. I have a rough goal behind it, that of figuring out how to involve anthroponomy in my life against the social process obfuscation called the "Anthropocene." That rough goal does not resolve my initial apathy. But the holding pattern of essaying autonomy and disagreement in community with an eye toward the goal of explaining anthroponomy holds me over for a time. How, then, should I engage in my community's politics?

Notes

1 Clive Hamilton, *Requiem for a Species: Why We Resist the Truth About Climate Change*, New York: Routledge, 2015 and *Defiant Earth: The Fate of Humans in the Anthropocene*, Malden, MA: Polity Press, 2017; Naomi Oreskes and Erik Conway, *The Collapse of Western Civilization: A View from the Future*, New York: Columbia University Press, 2014; Roy Scranton, *Learning to Die in the Anthropocene: Reflections on the End of a Civilization*, San Francisco: City Lights Books, 2015; Rupert Read, "This Civilization Is Finished. So What Is to Be Done?" *Shed a Light* series, Churchill College, University of Cambridge, November 7, 2018.
2 Cf. "Civilization," *Oxford American Dictionary*, Apple Inc., 2005–2017.
3 Cf. Neil Brenner, *New Urban Spaces: Urban Theory and the Scale Question*, New York: Oxford University Press, 2019, chapters 9–10. See especially figures 10.9–10.10.
4 On the colonial world system, see Jan C. Jensen and Jürgen Osterhammel, *Decolonization: A Short History*, translated by Jeremiah Riemer, Princeton: Princeton University Press, 2017. On colonial terms of self-determination verses indigenous terms – and forms of political organization – see Glenn Coulthard, *Red Skin, White Masks: Rejecting the Colonial Politics of Recognition*, Minneapolis: University of Minnesota Press, 2014. See also the emerging work of Kyle Powys Whyte on "seasonal sovereignties," e.g. "Food Justice and Collective Food Relations," in A. Barnhill, M. Budolfson and T. Doggett, eds., *Food, Ethics and Society: An Introductory Text with Readings*, New York: Oxford University Press, 2016, pp. 122–134.
5 See Ta-Nehisi Coates, *Between the World and Me*, New York: Spiegel & Grau, 2015; Martin Luther King, Jr., *Where Do We Go from Here? Chaos or Community*, foreword by Coretta Scott King, Boston: Beacon Press, 2010. For an accessible reflection on the "neocolonial," see Falguni A. Sheth, "Neocolonialism's Anxieties: Schooling Women of Color," *Blog of the APA*, July 31, 2019.
6 This point is adapted from Kyle Powys Whyte, "Indigenous Science (Fiction) and the Anthropocene: Ancestral Dystopias and Fantasies of Climate Change Crises," *Environment and Planning E: Nature and Space*, v. 1, n. 1–2, pp. 224–242.
7 Cf. Extinction Rebellion, "The Truth," accessed September 18, 2019.
8 On the international order: Stephen M. Gardiner, *A Perfect Moral Storm: The Ethical Tragedy of Climate Change*, New York: Oxford University Press, 2011; on the global economy: Geoff Mann and Joel Wainwright, *Climate Leviathan: A Political Theory for Our Planetary Future*, Brooklyn, NY: Verso, 2018.
9 Mann and Wainwright, 2018; cf. John Dryzek and Jonathan Pickering, *The Politics of the Anthropocene*, New York: Oxford University Press, 2019 on path dependencies, too.
10 See Immanuel Kant, *Critique of Pure Reason*, translated by Norman Kemp Smith, New York: St. Martin's Press, 1965, Second Division, Chapter 2, "Third Antinomy."
11 Kant called this awareness a "fact of reason." See Susan Neiman, *Moral Clarity: A Guide for Grown-Up Idealists*, New York: HarperCollins, 2008, chapter 3. Compare also her *Evil in Modern Thought: An Alternative History of Philosophy*, Princeton: Princeton University Press, 2015.
12 See my "Le citoyen apathique: quand les institutions étatique ne reflètent pas les droits de l'homme avec cohérence," *Annuaire Français de Relations Internationales*, Paris: Centre Thucydide, 2006, pp. 39–58.
13 Cf. Bernard Williams, *Shame and Necessity*, 2nd ed., Berkeley: University of California Press, 2008; the Greek heroes Williams studies could not look others in the eye, because they had failed the expectations of someone they admired so much their world

was built around it ("Necessary Identities," p. 103ff). I speak deliberately of not being able to look *oneself* in the face, as in a mirror – a trope of autonomy.

14 Stephen Gardiner, "Accepting Collective Responsibility for the Future," *Journal of Practical Ethics*, v. 5, n. 1, 2017, pp. 22–52, esp. p. 33: "Instead, on the view I have in mind, agency itself has its price. A heavy part of this price is the assumption of agency responsibility, the sort of moral and political responsibility that one has simply in virtue of being a moral agent in the first place."

15 On the quite literal social construction of a world, including its environment, see Steven Vogel, *Thinking Like a Mall: Environmental Philosophy After the End of Nature*, Cambridge, MA: MIT Press, 2015. The simple demands of Extinction Rebellion, 2019, are a start: facing up to climate reality, decarbonizing the economy; subjecting political processes to citizen's assemblies. It may even be that their demands – especially the first and third – as quintessentially anthroponomous.

16 Heidegger first saw the point about anxiety's relation to the contingency of norms, developing an ontological reading of Kierkegaard's focus on coming to terms with the contingency of one's existence. On freedom in this sense, see Jean-Luc Nancy, *The Experience of Freedom*, translated by Bridget McDonald, Palo Alto: Stanford University Press, 1994. For a refinement of Heidegger that brings out clearly the role norms play in his understanding of "existence," see John Haugeland, *Dasein Disclosed: John Haugeland's Heidegger*, edited by Joseph Rouse, Cambridge, MA: Harvard University Press, 2013. For the political extension of the appearance of the contingency of norms in the possibility of reorganizing society, see Jacques Rancière, *Disagreement: Politics and Philosophy*, translated by Julie Rose, Minneapolis: University of Minnesota Press, 2004.

17 Cf. Jonathan Lear, *A Case for Irony*, Cambridge, MA: Harvard University Press, 2011.

18 Will Steffen, Jacques Grinevald, Paul Crutzen, and John McNeill, "The Anthropocene: Conceptual and Historical Perspectives," *Philosophical Transactions: Mathematical, Physical and Engineering Sciences*, v. 369, n. 1938 and *The Anthropocene: A New Epoch of Geological Time?* March 13, 2011, pp. 842–867; Simon L. Lewis and Mark A. Maslin, "Defining the Anthropocene," *Nature*, v. 519, March 12, 2015, pp. 171–180; Will Steffan, Wendy Broadgate, Lisa Deutsch, Owen Gaffney, and Cornelia Ludwig, "The Trajectory of the Anthropocene: The Great Acceleration," *The Anthropocene Review*, v. 2, n. 1, 2015, pp. 81–98. On reorganizing the "rules of life," see Jeremy Bendik-Keymer and Chris Haufe, "Anthropogenic Mass Extinction: The Science, the Ethics, the Civics," in Stephen Gardiner and Allen Thompson, eds., *The Oxford Handbook of Environmental Ethics*, New York: Oxford University Press, 2017, Chapter 36.

19 See Scranton, 2015; Hamilton, 2017, among many others. For a pronouncement closer to the specifically political indignation of this chapter, see Mann and Wainwright, 2018, p. x.

20 The comparisons in this sentence refer to the main causes of several of the mass extinctions since life began on Earth, in particular, the end Cretaceous extinction and the end Permian extinction. For an early version of this view of a planetary trauma, see James Lovelock, *The Revenge of Gaia: Earth's Climate Crisis & the Fate of Humanity*, New York: Basic Books, 2007. Perhaps more relevant to this book, cf. his *A Rough Ride to the Future*, New York: The Overlook Press, 2015.

21 Dale Jamieson, *Reason in a Dark Time: Why the Struggle Against Climate Change Failed – and What It Means for Our Future*, New York: Oxford University Press, 2014; see also Dale Jamieson, "Ethics, Public Policy, and Global Warming," in Allen Thompson and Jeremy Bendik-Keymer, eds., *Ethical Adaptation to Climate Change: Human Virtues of the Future*, Cambridge, MA: MIT Press, 2012, pp. 187–202 – a revised version of Jamieson's seminal 1992 article that may have been the first to make this point about responsibility, long before Peter Singer's *One World: The Ethics of Globalization*, New Haven: Yale University Press, 2002, Chapter 2, "One Atmosphere."

Finally, see Allen Thompson, "The Virtue of Responsibility for the Global Climate," in Thompson and Bendik-Keymer, eds., 2012, pp. 203–222; Gardiner, 2017.

22 The record, for instance, concerns decarbonizing the economy. On the world system's failures to do that so far, projecting unrealistic promises into this century when global warming begins to be well far gone for the world's vulnerable populations of humans and more than humans, see Jamieson, 2014; Mann and Wainwright, 2018. The latter has a good selection of IPCC and post-Paris Accord sources, including their contradictions. The former tells the story of the industrial world's failure to head off dangerous global warming in the last decades of the twentieth century.

23 Cf. Tom Cohen, "Trolling *Anthropos* [*sic*.] – or Requiem for a Failed *Prosopopeia* [*sic*.]," in Tom Cohen, Clare Colebrook, and J. Hillis Miller, eds., *Twilight of the Anthropocene Idols*, London: Open Humanities Press, 2016, pp. 20–80; Clare Colebrook, "What Is the Anthropo-Political?" in Cohen, Colebrook, and Miller, 2016, pp. 81–125.

24 In the preface to their collectively written volume, Cohen, Colebrook, and Miller imply that the rhetorical purpose of specifying causality for our planetary situation is to wash one's hands of blame; Cohen, Colebrook, and Miller, 2016, pp. 7–8. That's an illicit conclusion. Getting causality clear is accurate. What one does *with* accuracy is a different issue. It's subject, for instance, to moral norms (e.g., of non-avoidance of one's responsibilities). When the editors go on to say that specifying causal narratives cover over a question about narrative itself, they again jump to conclusions. Narrative may always depend on ignoring some details and some limits. But narrative in the service of accuracy is open to *coming to terms with* its own ignorance. To see narrative as in the service of accuracy is a condition of criticism, not its repression.

25 Mann and Wainwright, 2018.

26 See especially Martin Gorke, *The Death of Our Planet's Species: A Challenge to Ecology and Ethics*, translated by Patricia Nevers, Washington, DC: Island Press, 2003.

27 Not only *capitalism* looms large here, although it *mainly* does. "Modern" production also does, or as I would say, industrial production. See Moishe Postone, *Time, Labor, and Social Domination: A Reinterpretation of Marx's Critical Theory*, New York: Cambridge University Press, 1998. See also Kyle Powys Whyte, "Indigenous Climate Change Studies: Indigenizing Futures, Decolonizing the Anthropocene," *English Language Notes*, v. 54, n. 1–2, Fall 2017a, pp. 153–162.

28 See Whyte, 2017a, p. 159: "A term like 'anthropogenic' has very diverse meanings for Indigenous peoples, from gradual changes, such as the adoption of new 'relatives' (e.g., adoption of the horse in North America) to the shaping of habitats for certain plants and animals, to disruptive settler colonialism, such as practiced by Europeans arriving in North America. 'Anthropogenic climate change' or 'the Anthropocene,' then, *are not precise enough terms* for many Indigenous peoples, because they *sound like all humans are implicated in and affected by colonialism, capitalism and industrialization in the same ways*" (emphasis mine). The shadow of colonialism would then be to use a term that sweeps away distinctions in forms of society such that the *vicious* forms are lumped together with more sustainable ones, making colonial society's forms speak for all.

29 Kyle Powys Whyte, "Our Ancestors' Dystopia Now: Indigenous Conversation and the Anthropocene," in Ursula Heise, Jon Christensen, and Michelle Niemann, eds., *Routledge Companion to the Environmental Humanities*, New York: Routledge, 2017b, pp. 2016–2015.

30 Kent G. Lightfoot, Lee M. Panich, Tsim D. Schneider, and Sara L. Gonzalez, "European Colonialism and the Anthropocene: A View from the Pacific Coast of North America," *Anthropocene*, 4, 2013, pp. 101–115; Heather Davis and Zoe Todd, "On the Importance of a Date, or Decolonizing the Anthropocene," *ACME: An International Journal for Critical Geographies*, v. 16, n. 4, 2017, pp. 761–780. Finally, see also Whyte, 2017a.

31 Cf. "Primitive Accumulation," Karl Marx, *Capital, v.1: A Critique of Political Economy*, translated by Ben Fowkes, New York: Penguin Books, 1992, Chapter 31.

32 The spatial logic of colonialism takes the land as a resource contained by violent expropriative sovereignty; it does not see the land as living, wherein people form a relationship with it expanding the possibilities of kin. See Coulthard, 2014, esp. chapter 2 and its notion of "grounded normativity." See my *The Ecological Life: Discovering Citizenship and a Sense of Humanity*, Lanham, MD: Rowman & Littlefield, 2006, lectures 2, 4, and 5 especially. See Mann and Wainwright, 2018, p. 196, which also cites Coulthard, 2014, p. 60. Finally, see the work of Shiri Pasternak on behalf of the Barriere Lake Algonquins, *Grounded Authority: The Algonquins of Barriere Lake Against the State*, Minneapolis: University of Minnesota Press, 2017.

33 I find it so strange that scientists and scholars have pinned the "Anthropocene" on all of humankind when obviously specific social processes have driven the worst problems we now face that I end up doubting their accountability. It is not that the scientists are corrupt, but that the institutional system around science may be, in so far as it blinkers its practitioners to disregard the nuances of social studies (Mann and Wainwright, 2018, chapter 3), shaped in part by the way that scientists are funded to support the economy, not question it. See Steven Shapin, *The Scientific Life: A Moral History of a Late Modern Vocation*, Chicago: University of Chicago Press, 2009, chapter 7, "The scientific entrepreneur."

Of the humanists making waves by using the "Anthropocene," I am more skeptical, since humanists are trained to be nuanced about representational claims. The "Anthropocene" is a word representing humankind as such. But given the decreasing interest in the humanities within the global economy, it wouldn't be surprising if humanists capitalized on a moment of apparent grand relevance and claimed the stage. See Martha C. Nussbaum, *Not for Profit: Why Democracy Needs the Humanities*, Princeton: Princeton University Press, 2010 or Wendy Brown, *Undoing the Demos: Neoliberalism's Stealth Revolution*, Cambridge, MA: Zone Books, 2015, among many other books.

34 For some popular writing on neoliberalism and its culture of avoiding accountability, which I have called "arbitrarianism" in the context of the United Stated of America, see my "This Conversation Never Happened," *Tikkun*, March 26, 2018. Note the role of self-ownership in this story. It and its Lockean legacy will be important for this book's wresting of autonomy away from liberty.

35 Dipesh Chakrabarty, "The Climate of History: Four Theses," *Critical Inquiry*, v. 35, n. 3, 2009, pp. 197–222.

36 On the reification of "nature," which is most appropriate here since social processes are turned into a biogeological force of which we were unaware, see Vogel, 2015, especially chapter 3; for a history of reification in critical conversation with Lukács, see Axel Honneth, *Reification: A New Look at an Old Idea*, with Judith Butler, Raymond Guess, and Jonathan Lear, edited by Martin Jay, New York: Oxford University Press, 2008.

37 Georg Lukács, "Reification and the Consciousness of the Proletariat," in Rodney Livingstone, trans., *History and Class Consciousness*, Cambridge, MA: MIT Press, 1971, pp. 83–222, p. 83, cited in Honneth, 2008, p. 21.

38 See Vogel, 2015 again, also his "Alienation and the Commons," in Thompson and Bendik-Keymer, 2012, chapter 14.

39 Cf. Kristie Dotson, "Tracking Epistemic Violence, Tracking Practices of Silencing," *Hypatia*, v. 26, n. 2, 2011, pp. 236–257. Thus, e.g., a scientist may say that the cause of our situation is humankind, thereby silencing the testimony of indigenous people that their relation to land would never permit such a violence as has been caused by colonialism scaled up into industrial and state-based world systems.

40 Karl Marx, *Economic and Philosophic Manuscripts of 1844* in Karl Marx and Friedrich Engels, *Economic and Philosophic Manuscripts of 1844* and *The Communist*

Manifesto, translated by Martin Milligan, Amherst, NY: Prometheus Books, 1988, pp. 13–170. See also Vogel, 2015, chapter 3.

41 Jean-Paul Sartre, *Being and Nothingness*, translated by Hazel E. Barnes, New York: Washington Square Press, 1984, part one, chapter 2. The root of this distinction can be found in Kant, 1965, Second Division, chapter 2, "Third Antinomy."

42 The concept of autonomy has come under criticism in social philosophy for about half a century, including recently in some environmental thought and in some decolonial philosophy. The criticism often depends on a straw-man, a simplified notion of autonomy. But protecting the concept of autonomy from simplification is important for justice. One mistake many criticisms of autonomy make is to confuse autonomy with an ontological version of *independence*. This, I believe, is actually a conflation of a Lockean inheritance of *liberty* (of freedom from interference) with the intrinsically relational structure of autonomy. However, to conceptualize the depth to which we are interdependent – or in what Timothy Morton calls the "mesh" – is not to undermine autonomy. Beings in relationship can be autonomous, and autonomy involves acknowledging our dependence so that we may figure out what makes sense to us in life and can be justified as a way of being. On this last point, see Luce Irigaray, *To Be Two*, translated by Monique M. Rhodes and Marco Cocito-Monoc, New York: Routledge, 2001, chapters 6–8, 10. Cf. Timothy Morton, *Ecology Without Nature*, Cambridge, MA: Harvard University Press, 2009. On prominent critical frameworks against autonomy, see Jane Bennett, *Vibrant Matter: A Political Ecology of Things*, Durham, NC: Duke University Press, 2010; Fred Moten, *consent not to be a single being – v. 3: The Universal Machine*, Durham, NC: Duke University Press, 2018.

43 On authority, cf. Jean Hampton, *The Authority of Reason*, New York: Cambridge University Press, 1998, chapters 2–3; on intention, see G.E.M. Anscombe, *Intention*, 2nd ed., Cambridge, MA: Harvard University Press, 2000; on the indignity of being pushed about, see Martha C. Nussbaum, *Women and Human Development: The Capabilities Approach*, New York: Cambridge University Press, 2000, chapter 1 (the explanation of the *spirit* behind the entire capability approach); finally, on all of these points, compare Christine Korsgaard, *Self-Constitution: Agency, Identity, and Integrity*, New York: Oxford University Press, 2009.

44 Part of the subtlety followed in this paragraph is the antinomy between freedom and determinism isolated by Kant, 1965. Morality is a point of view, one necessary to living a recognizably human life with relations of care, accountability, and the like. In Wittgensteinian terms, the "grammar" of human life depends on the moral point of view, which we take up *against* realizations of our having been pushed around and unable to live out the "fact of reason" in constructing a world that makes sense to us, i.e., is autonomous. See my 2006 for other uses of the notion of Wittgensteinian grammar in environmentalism.

45 See Gardiner, 2011, again.

46 I first used this term in "Social process obfuscation and the anthroponomy criterion," *Earth System Governance Project* annual meeting, Utrecht University, November 2018.

47 Cf. Whyte, 2017a.

48 Manuel Castells, *The Rise of the Network Society*, 2nd ed., Cambridge, MA: Blackwell, 2000.

49 Brown, 2015; cf. David Harvey, *A Brief History of Neoliberalism*, New York: Oxford University Press, 2007; cf. Gardiner, 2017.

50 Coulthard, 2014, chapter 1.

51 Gardiner, 2011.

52 This was an implication of my first environmental study, *The Ecological Life*, 2006. The main moral system I examined there was the one articulated in the overlapping consensus of the *Universal Declaration of Human Rights*. This document has wide acceptance across nations and the major world religions. I also considered the

widespread acceptance of respect for life as a religious, cultural, or artistic commitment. On this last point, see also my "The Sixth Mass Extinction Is Caused by Us," in Thompson and Bendik-Keymer, 2012, pp. 263–280 and "Species Extinction and the Vice of Thoughtlessness: The Importance of Spiritual Exercises for Learning Virtue," in Philip Cafaro and Ronald Sandler, eds., *Virtue Ethics and the Environment*, New York: Springer, 2010, pp. 61–83. Finally, the problem of self-contradiction is taken up in my "Living Up to Our Humanity: The Elevated Extinction Rate Event and What It Says About Us," *Ethics, Policy & Environment*, v. 17, n. 3, 2014, pp. 339–354. See also Gardiner, 2017's picking up of the language of the "gap" between institutions and commitments which also appears in "The Sixth Mass Extinction Is Caused by Us."

53 Coulthard, 2014, chapter 2, and Mann and Wainwright, chapter 8.

54 Mann and Wainwright, 2018, pp. 35–38 on the Paris Agreement as a case in point; also, chapters 7–8. Yet for a sense of how complex the architecture of inclusion of agents is, and of how difficult it can be to move the public sphere to the place where it is powerful with world system agencies such as states and international governance bodies, see Frank Biermann, *Earth System Governance: World Politics in the Anthropocene*, Cambridge, MA: MIT Press, 2014, especially chapters 3–5 and 7; and Hayley Stevenson and John Dryzek, *Democratizing Global Climate Governance*, New York: Cambridge University Press, 2014, especially chapters 6–8.

55 For a theory that spans both capitalist and non-capitalist industrial economies, see Postone, 1998.

56 The critique of what is taken to be "realistic" in the sense of "pragmatic" is a theme in Jacques Rancière's work where he develops the Rousseauian tradition of examining how educational and political "realisms" run cover for hegemonic and unjust orders. See Jacques Rancière, *The Ignorant Schoolmaster: Five Lessons in Intellectual Emancipation*, translated by Kristin Ross, Palo Alto: Stanford University Press, 1991, and Rancière, 2004. See also the discourse of "adaptive preferences" in the Capability Approach: Nussbaum, 2000; Serene J. Khader, *Adaptive Preferences and Women's Empowerment*, New York: Oxford University Press, 2011.

57 In commentary on a draft, Joel Wainwright asked me what my view of ideology is. I took him to be suggesting that the avoidance of contradictions around the "Anthropocene" is ideological, even hegemonic, in Gramsci's sense. Later in this book, I will discuss the jargon word "coloniality" and its related proposal to name an ideology of ongoing imperialism that maintains the power structures and privileges, also the inequalities, of the colonial world order, whether decolonized or still awaiting settler decolonization. What I've assumed here is perhaps more modest than ideology.

58 Cf. Byron Williston, *The Anthroponomy Project: Virtue in the Age of Climate Change*, New York: Oxford University Press, 2015. Williston's project is sympathetic to my own in some respects, namely in his attempt to make the "Anthropocene" a matter primarily of prospective responsibility for producing a morally acceptable future. Yet his use of the name repeats social process obfuscation, and his way of conceptualizing responsibility through the virtues of truthfulness and justice, among others, fails to create a concept for the social construction of the political order our planetary situation morally demands. I argue that the concept should be anthroponomy. On other appeals to virtue, see Thompson and Bendik-Keymer, 2012. In my contribution to that volume, I suggested that virtue must be understood as an aspect of civic engagement and political responsibility and gestured to both social structures and areas of ignorance that push action "upward" to collective action, away from merely personal virtue. Vogel's contribution also did this directly. On a partial landing consistent with that upward push, see Gardiner, 2017.

59 See Neil Brenner, "Is the World Urban?" Vincent Scully Lecture, University of Miami School of Architecture, April 2, 2018, accessed December 26, 2018, especially his opening comments on reflexivity, theory and social reality. See also Gardiner, 2011, especially "The Theoretical Storm."

60 The notion of reason making demands on the world comes from Kant's understanding of theory as a guide to the construction of a world we can accept as reasonable. It is part of his adaptation of Rousseau's general will. See Susan Neiman, *The Unity of Reason: Rereading Kant*, New York: Oxford University Press, 1994.

61 As I noted in a footnote in the preface, I first used anthroponomy in "Ethical Adaptation to Climate Change," *European Financial Review*, October–November 2012, pp. 22–26. The most articulated paper-length introduction of it is in " 'Goodness Itself Must Change' – Anthroponomy in an Age of Socially-Caused, Planetary, Environmental Change," *Ethics & Bioethics in Central Europe*, v. 6, n. 3–4, 2016, pp. 187–202. My first discussion of anthroponomy in relation to decolonization was in "Decolonialism and Democracy: On the Most Painful Challenges to Anthroponomy," *Inhabiting the Anthropocene*, July 27, 2016. To my knowledge, there is no other philosophical use of the term "anthroponomy" except in Kant's *Metaphysics of Morals*. See, for instance, David Susskind, *The Idea of Humanity: Anthropology and Anthroponomy in Kant's Ethics*, New York: Routledge, 2015. An online circulating use of "anthroponomy" as a name given to human-environment interactions in a sub field of anthropology has, however, been unsubstantiated in my research with anthropologists who specialize in the area where the sub-field should be. Finally, I found one use of "anthroponomy" in Gorke, 2003, to name the giving of value to nature by humans. But the word for this would appear to be more precisely "anthroponoēsis."

62 See especially my 2016 article.

63 When something isn't true or goes against people's core beliefs, it isn't justifiable.

64 Cf. Iris Marion Young, *Responsibility for Justice*, with a foreword by Martha C. Nussbaum, New York: Oxford University Press, 2011; Thompson, 2012; compare Gardiner, 2017.

65 This justifiability of effects only in the eyes of the affected was taught me by Amy Linch in her unpublished work on the capability approach, other species, and Deweyan democracy.

66 Cf. Philip Pettit, *Republicanism: A Theory of Freedom and Government*, New York: Oxford University Press, 1997.

67 Cf. Coulthard, 2014, on "practices of self-recognition" in "self-determination," especially chapter 5.

68 Gardiner, 2011; Jamieson, 2014; Mann and Wainwright, 2018.

69 Henry Shue, "Deadly Delays, Saving Opportunities: Creating a More Dangerous World?" in Stephen M. Gardiner, Simon Caney, Dale Jamieson and Henry Shue, eds., *Climate Ethics: Essential Readings*, New York: Oxford University Press, 2010, pp. 146–162.

70 Gardiner, 2011, p. 145ff.

71 Gardiner, 2011, p. 145ff.

72 Gardiner, 2011, p. 145ff, chapter 5, "Intergenerational Buck-Passing."

73 Cf. Gardiner, 2017.

74 See my "Presentism the Magnifier," IAEP/ISEE, University of East Anglia, June 12–14, 2013.

75 See Michel Foucault, *Abnormal: Lectures at the Collège du France, 1974–1975*, translated by Graham Burchell, New York: Picador, 2004, *Discipline and Punish: The Birth of the Prison*, translated by Alan Sheridan, New York: Vintage Books, 1995 and *The History of Sexuality, v. 1: An Introduction*, translated by Robert Hurley, New York: Vintage Books, 1990; also, Lynne Huffer, *Mad for Foucault: Rethinking the Foundations of Queer Theory*, New York: Columbia University Press, 2009; Moten, 2018.

76 Cf. Giorgio Agamben, *The Open: Man and Animal*, translated by Kevin Attell, Palo Alto: Stanford University Press, 2003.

77 See Cohen, Colebrook and Miller, 2016, "Preface," and Colebrook, 2016, where the suspicion and suggestion arise that the Anthropocene is the pretext for consolidating

all of humankind into a unit. For these authors, the Anthropocene *makes* humankind at the moment humankind appears at risk.

78 "Humanity" is a virtue or a quality of character. "Humankind" is a kind of being. See *The Ecological Life*, 2006, lecture 4, and my *Conscience and Humanity*, dissertation submitted to the University of Chicago Department of Philosophy, 2002, chapter 2; see also, Theodore Zeldin, *An Intimate History of Humanity*, New York: Harper Perennial, 1995.

79 See especially here Gayatri Chakravorty's discussion around "planetarity," in " 'Planetarity' Box 4 *(Welt),*" *Paragraph*, v. 38, n. 2, 2015, pp. 290–292 and her *Death of a Discipline*, New York: Columbia University Press, 2005, chapter 3. See also her general concern in "Can the Subaltern Speak?" in Rosalind Morris, ed., *Can the Subaltern Speak? Reflections on the History of an Idea*, New York: Columbia University Press, 2010, part I.

80 Cf. Moten, 2018; Rancière, 2004; also, Foucault, 2004; Huffer, 2009; see also Alain Badiou, Pierre Bourdieu, Judith Butler, Georges Didi-Huberman, Sadri Khiari, and Jacques Rancière, *What Is a People?* translated by Jody Gladding, New York: Columbia University Press, 2016 and – for its clarifications of the problems of identification– Jacques Rancière, "Critical Questions for the Theory of Recognition," in Katia Genel and Jean-Philippe Deranty, eds., *Recognition or Disagreement: A Critical Encounter on the Politics of Freedom, Equality, and Identity*, New York: Columbia University Press, 2016, pp. 83–95. Finally, consider the point when considering Fred Moten, *consent not to be a single being, v. 1: Black and Blur*, Durham, NC: Duke University Press, 2017.

81 Rancière, 2004, chapter 2, "Wrong."

82 Cf. *The Ecological Life*, 2006, lecture 8.

83 Foucault rightly remains suspicious of pat answers to the problem of "soft" domination, the supple chains that bind us by being our identities. But see how much he does this just to keep open disagreement in his "What Is Critique?" in James Schmidt, ed., *What Is Enlightenment? Eighteenth Century Answers and Twentieth Century Questions*, Berkeley: University of California Press, 1996, pp. 382–398.

84 Percy Bysshe Shelley, "Ozymandias," 1818.

85 Cf. Coulthard, 2014.

86 Cf. Kristie Dotson, "Making Sense: The Multistability of Oppression and the Importance of Intersectionality," in Namita Goswami, Maeve M. O'Donovan, and Lisa Yount, eds., *Why Race and Gender Still Matter: An Intersectional Approach*, London: Pickering & Chatto, 2014, pp. 43–57.

87 Ludwig Wittgenstein, *Philosophical Investigations*, translated by G.E.M. Anscombe, P.M.S. Hacker and Joachim Schulte, 4th ed., Malden, MA: Blackwell, 2009, I.67.

88 *Oxford American Dictionary*, 2005–2017, "Autonomy."

89 Cf. Michel Foucault, *The Government of Self and Others: Lectures at the Collège du France, 1982–1983*, translated by Graham Burchell, New York: Picador, 2011; also see Foucault, 1996.

90 See Nussbaum, 2000, pp. 136–144, 164–165, 307. But also Rosa Terlazzo, "The Perfectionism of Nussbaum's Adaptive Preferences," *Journal of Global Ethics*, v. 10, n. 2, 2014, pp. 183–198 and "Conceptualizing Adaptive Preferences Respectfully: An Indirectly Substantive Account," *The Journal of Political Philosophy*, v. 24, n. 1, 2016, pp. 206–236.

91 Cf. Immanuel Kant, *Groundwork of the Metaphysics of Morals*, translated by Mary Gregor and Jens Timmerman, 2nd ed., New York: Cambridge University Press, 2012; my formulation is not Kant's in many respects, starting with my emphasis on being open to challenge in a real or hypothetical confrontation, rather than with Kant's emphasis on the form of the moral law. Still, the legacy is Kantian. As with Martha C. Nussbaum, I view more than human animals as morally considerable and as capable of making tacit moral demands on us, given our translation of their obvious

behavior and body language; also, of their being deserving of trusteeship in many instances. See Martha C. Nussbaum, *The Frontiers of Justice: Disability, Nationality, Species Membership*, Cambridge, MA: Harvard University Press, 2006, and my "From Humans to All of Life: Nussbaum's Transformation of Dignity," in Flavio Comim and Martha C. Nussbaum, eds., *Capability, Gender, Equality: Toward Fundamental Entitlements*, New York: Cambridge University Press, 2014, pp. 175–192. On being open to justification before all, cf., Emmanuel Levinas, *Otherwise Than Being or Beyond Essence*, translated by Alphonso Lingis, Dordrecht, Netherlands: Kluwer Academic Publishers, 1991.

What I am claiming about autonomy does not fit neatly into the dismissal Mann and Wainwright, 2018, make of philosophical approaches to interpersonal accountability in politics, thereby apparently missing the important points about speech and openness to claims *against* one's norms made by Rancière, 2004, chapters 1–3. It seems to me that Rancière understands that autonomy is not about norms set up *in advance* of people's outrage, thereby meant to police people's speech, but is about *being ready in the moment* to come to account for an outrage that suddenly appears in clear day in the speech or communication of people or morally considerable beings who are beside themselves. It is about being accountable to suspending the norms one had thought unobjectionable so as to come to account for a wrong shown to exist in the institutional and normative structure and fabric of one's society. Autonomy is accountable for reorganizing society in light of an inclusion that cannot be predicted or foreclosed in advance.

92 Cf. Gardiner, 2017.
93 Autonomy is a *condition* of a life that successfully makes sense to oneself over time through disagreement. It is a *power* of being able to judge what makes sense, consider it with others through reasoning and sense-making, and live according to the determinations arrived therethrough. It is a *process* of squaring up with sense, disagreeing with what one has absorbed, responding to disagreement from others, and adjusting or even transforming one's life to make better sense.

As I previously mentioned, it has become popular among some theorists to point to the many ways that we are not autonomous. But the result of showing the ways that we are not autonomous, aren't ourselves, are dependent, related, un-discrete, or shaped in our minds and bodies is that awareness of them helps us refine what makes sense. It makes us more autonomous. In criticizing autonomy, we become more aware of who we are, independent when we actually can be, discrete in our relations so that they are focused in their intimacy, trust, and safety; and giving ourselves the mental space to think about our lives otherwise so that the norms we follow are the ones by which we actually *abide*. See Foucault, 1996, especially the notion of a "critical attitude" of "space" around norms. Interestingly, Foucault is no foe of autonomy in this essay – the same Foucault whose word on abnormality, for instance, is often seen to present problems precluding autonomy. See Foucault, 2004; Huffer, 2009.

Figure 2.1 Old room, New Hartford, New York, 1986

2 How should I relate to colonialism?

**Later in the summer, Shaker Heights, Ohio,
once land of many nations**

Whether out of half-dream or the world's stillness I don't know, I woke with words more or less in my mind: "Do we have a relation that is morally acceptable?" The sentence was there in entirety, but it was unclear – a relation to what? When I thought for a moment, I knew. I was referring to our planetary situation, about which I'd been anxious over a month ago in those ten days of focused writing.

Why a relation? The expression was vague and odd, since we have many possible relationships. I seemed to be searching for a single basic form to hold things steady. Is anthroponomy a single, basic form?

The other thing that was odd about the dream fragment was that it was equivocal. It seemed that I was asking myself both whether I knew of any relation and whether that relation, whatever it is, were actual. This ambiguity fit where I was last month, for I remain unconvinced that "anthroponomy" can actually guide my action. Can "anthroponomy" be a clear relation that I can actually develop?

Before this thought finished, however, another sentence sprung to mind: "And if we do not,[1] how should I engage in my community?"

Once again, this question left open "anthroponomy" as a possible answer. But I need to know what anthroponomy is in a relatable way.

*

So on a summer morning, I reentered my community from oblivion. If we are to find a morally acceptable relation to our planetary situation, it must make sense to us, plunging into our oblivion – the world inside us from which sense emerges that is most intimate to us, subject then to critical reflection on whether it makes good sense. Legitimacy isn't abstract. It goes to the core of oneself – to the reflexive practice by which one sticks with what one believes, follows out what one intends, acknowledges what one desires and more generally feels.[2] If legitimacy does not go into our oblivion, a sunken part of ourselves will resist the norm, authority, or common sense in question[3] and, by implication, submit to it insofar as we repress our qualms to wear a stolid face.

In a sense, community begins in our oblivion. For there can be no community of ours that excludes ourselves, even by internal self-repression. A collection of people self-repressing isn't a community. We need each other, open, for a community to be communal.[4]

Is oblivion needed *mutatis mutandis* for autonomy? For anthroponomy? Community, autonomy, and anthroponomy, it would seem, share the common logical condition of including ourselves in them. How could I authorize anything to govern myself if I did not believe in it deep down? How could I participate in the reflexive regulation of humankind if I were excluded from participating as a fellow human? So, yes. Not one of the three – community, autonomy, or anthroponomy – accepts submission in place of true legitimacy.

An odd place to begin with politics – on the edge of sleep.

An odder place to begin with collective responsibility for a planetary situation – on the edge of oblivion.[5]

*

I am trying to understand prospective responsibility for our planetary situation in a way that maintains a commitment to autonomy and, particularly, the non-domination of anyone. I am trying to understand my moral relation to our planetary situation locally – in my life and community – so that I can see how the form of a morally sufficient response could be relatable.

I know, too, that in a society involving injustice and wantonness, a life of responsibility, autonomy and of non-domination "is not something that you find; it is something that you make."[6] With the lack of anthroponomy in the world, I will have to use critical thinking to get to the bottom of what it should, or must, involve. My hope is that there will be connections that slowly emerge, translatable forms, that allow me to understand the intuition that anthroponomy may be "fractal" across scales and places, and that I will be able us involve it in my day-to-day living and political engagement.[7]

Two weeks later, Shaker Heights, Ohio

I was out running this morning when I saw something that appalled and perplexed me. It was a flag of the United States of America, only with the stars missing and, in their place, the caricature of a Native American, "Chief Wahoo," the made up "Indian" that serves as the mascot of the Major League Baseball team in my city – the "Cleveland Indians."[8]

The road where I saw it – one adjacent to Woodlawn Avenue, I think – is a quiet, propertied road. Of solid brick and stone and with slate roofs, the houses rise from lawns spread widely about them, meticulously kept. Some houses are mansions, the rest refined or quaint.

Although Shaker Heights is one of the most racially diverse areas in the Cleveland metro-region where I live, the metro-region is one of the most segregated in the United States of America.[9] Whenever I jog through this area around Woodlawn Avenue, most of the people out in their yards are white. Native Americans are a fraction of a percent of the population of this small city contiguous with the city of Cleveland on the inside of the beltway.[10] For some "Americans," Shaker Heights is ideal: a good public-school system, safe neighborhoods, well-educated neighbors, classy and not gaudy digs.[11] "Establishment" might be the caption of a high school senior's photography exhibit of the area.

There, swaying innocuously in the establishment, the flag. It was faded – the red dull and the blue dirty, the white dusted up. Seemingly hung over many years during the spring, summer, and fall of the baseball season, the flag implied, "Go Cleveland!"

A strange patriotism peered from this flag. The Cleveland Indians and the United States were one, joined by the "red Sambo."[12] Is it okay in this community as a patriot to display a racist caricature of colonization? Does my community ignore the violence implied by what is okay? With complex symbolism billowing across its fabric, the flag erased colonialism's violent history and made domination seem normal.[13]

Imagine "Chief Wahoo" appearing plainly across the city each spring were the character to offend basic moral and legal norms of the community and oppose its historical sense. At the least, there would be disagreement, neighbors angry at neighbors, a visit from a police community relations officer called by a neighbor to see if neighborliness is a sufficient reason to remove the flag. Yet the flag is within the bounds of common sense here. How is that possible?

I wondered, was flying the flag a commonly acceptable expression of people's autonomy in my community? Presume that the people who flew the flag had the opportunity and space to think for themselves, thought that flying the flag made sense (e.g., if they found it humorous or witty, it did make sense to them), and thought that flying the flag was justifiable to others as something that falls within the norms of the society that the flag owners accept. Then flying the flag would be autonomous. It would be free expression emerging out of a space of liberty,[14] cohering with what the owners believe is true or good, and consistent with what they think can be justified to others with whom they live in a society whose general norms they accept. They would not be acting under constraint, against their beliefs, or in opposition to the things they can justify to others with whom they share the world. Barring coercion or some other clear deformation of their capacity to judge for themselves, their common sense would be autonomous.

Of course, the owners of the flag needn't have worked through all their personal, psychological issues or be completely without self-deception or absolutely clear of delusion. They needn't have no seductions, peer pressure, or tough spots in their lives that might give them motives to avoid thinking some things though or to stop scrutinizing other things. The issue here is more modest. Given a chance to think for themselves and to consider alternatives, did the flag-flyers act in a way that made sense to them and which they thought was reasonable? If so, they were not obeying something that rushed them to decide without adequate thought, didn't make sense to them, or seemed obviously unjustifiable in their community where they live. There's no obvious way that the flag-owners would have been heteronomous. Rather, they were – let's say – ordinarily autonomous.[15] They seemed to follow norms that they could affirm themselves, which they would give themselves, and which they find legitimate.

So a problem with autonomy appeared to me. To what extent and in what ways can autonomy enable racism, historical obliviousness or a history of domination hidden from view? This problem could make autonomy unsuitable for our planetary situation, for it could then normalize histories of injustice, something the "Anthropocene" does in a different way.

Some days later, Shaker Heights, Ohio

The racist flag marks a territory of confusion. It's well known in Cleveland that the mascot, "Chief Wahoo," is opposed by indigenous and social justice groups. The absence of disagreement in my community around the flying of flags that appear to perpetuate injustice makes me doubt that I understand how autonomy and community ought to relate or that I am missing some obstacle to their relation. How could something be plainly justifiable out in the open when it's plainly contested as a matter of human dignity?[16] Where is accountability to the dignity of others? What gives people the license – or should I say "liberty"?! – to avoid moral accountability to each other? If the common moral and legal norms where I live normalize historical injustice and make it easy for entitled people to ignore protests of indignity, how is where I live a "community"?

But what if autonomy didn't dismiss? In my community, there doesn't seem to be accountability to those who sincerely contest common sense. Disagreement isn't joined to accountability. Rather, it seems possible to ignore what is sincerely disagreeable to others without working through the issue together. In the eyes and hearts of protesters, the license to remain unaccountable permits the ongoing ideology of colonialism and hides issues that are beneath it. Yet doesn't it seem obvious, on reflection, that in an active community, protesters shouldn't be relegated to shouting at the sky?[17] They are trying to say something vitally important to them, and it is worth listening to each other as a community. Why would it be right to block out the cries of those who claim an indignity?

The "Chief Wahoo" flag makes me doubt common sense here in Shaker Heights. Yet I also think that autonomy is being misunderstood. I don't see how autonomy can exclude disagreement. After all, autonomy is joined with sense and self. Concerning the world and our lives in it, sense is open to the world's evidence and disruptions. To keep finding and making sense is to be accountable to the disruptions and the world's evidence, considering and responding to it. At the same time, being oneself, broadly understood, is a practice of committing oneself or being coherent with what makes sense to oneself.[18] As we exist in the world, being oneself is also a practice of living with commitments such that one remains accountable to oneself and to others.[19] Both sense and self are thus relational, not atomistic, and both are accountable.

Disagreement ought to bring out the accountability implicit in autonomy. After all, disagreement disrupts sense among people who exercise their sense of self,[20] and, when sense is disrupted for a person with a sense of self, they become accountable to themselves regarding what makes sense. The accountability thereby relates autonomy to active and open community with the ones who have opened the disagreement. They cannot be denied in their claims to make sense, for they claim evidence that ought to be considered. In this way, disagreement with people who value autonomy brings out a specific form of inclusion of the ones who disagree. Together, you all have to consider what is, really, common sense, and that makes what you take to make sense open to change between you. Your sense of common sense may change accordingly and, with it, a rudimentary form of community with the ones who disagree may begin to appear *as* disagreement.

There is a whole landscape in these remarks. As a first run through it, we can say, then, that autonomy appears to be related within itself to any community worth its name – one capable of holding disagreement and changing accordingly.[21] This isn't more liberalism. I see the liberty – or should I say the "license"? – around me in this capitalist society founded on the principles first laid out by classical liberals such as John Locke.[22] But such a basic form of autonomous community – a community of autonomous disagreement – is different than liberal freedom from interference.[23] Liberal freedom from interference merely puts up with disagreement and so it merely puts up with community. But if someone objects to what I think makes sense and does so by asking – or demanding – that I consider something I have overlooked, and I still want to remain autonomous, then it seems that I must look. To ignore the disagreement would be to deny making sense of new considerations that arise. So I'd seem to cease trying to make sense of things, and that would be to lose the accountability to myself.

Wouldn't that render me heteronomous, no longer accountable with what I find makes sense in the evidence and disruptions of the world? Suppose, too, that I look and have no answer to the challenge. I could not then be said to have a clear grasp of what makes sense to me now that new considerations have come to light. If I continued on my way as before and did not reflect my new uncertainly in an openness to learning and to changing, then I would no longer act autonomously. In both cases of ignoring and avoiding my own unclarity, rather than figuring out what makes sense in the world to me as I actually come to understand it, I would avoid the roil of sense in the world and thus, ironically, myself.

Autonomy truly doesn't seem to pass through avoided disagreement, and it doesn't appear to permit a live-and-let-live attitude in situations of disagreement. It's not liberty. It makes more sense.[24]

Imagine that I am flying the flag and you object to it. Perhaps you point out that it is racist and that our community does not believe that racism is acceptable. Perhaps you also point out that the flag covers over the reality of colonialism, which is ongoing in the United States of America. Finally, suppose you point out that both things – the racism and the colonialism – are painful to many Native Americans and offensive to others in that they, as focused through the caricature, imply that Native Americans are not equally civilized, reasonable, or dignified.

The act of making fun of Native Americans without widespread Native American consent and acceptance of the caricature implies a lack of accountability to Native Americans as moral equals. Suppose you say all these things, and I scoff at you and assert my right to say whatever I want on my own property so long as it does not endanger public safety!

The question of what is my right is a consideration that you will have to consider, yes, but my having dismissed your concerns out of hand shows that I am not trying to make sense of what others can show me of the world. I am denying your view without having considered it. By dismissing you, I am suppressing the consideration of sense. Thus I am actually not committed to making sense. Not being committed to making sense, I am again heteronomous. The life I am living isn't one that makes sense of the world as people bring the world to light. And so now the life I am living is arbitrary. Willful and capricious, I'm not autonomous in this area of my life. License is heteronomy.

This arbitrariness, a willful denial of shared sense and of accountability with both of our views about the world, seems to me to be the opposite of autonomy.[25] I can also put the point about losing accountability to myself this way: By willfully denying your considerations, I also deny accountability for my own view, which involves beliefs about what is true and evaluations about what is good and bad, right and wrong, and many more, rich evaluative judgements. At that very moment, I lose *the* world as I shrink back into *my* world. How much of *a* world will it be if I persist arbitrarily? But how can I be autonomous if I lose a sense of the world?

Suppose, though, I did address your concerns and said that the caricature is neither racist nor colonialist. I would have to explain why it is neither. I would have to point to something that made sense. Suppose I said that the "Chief Wahoo" caricature doesn't show any feelings of superiority over Native Americans; it doesn't show any prejudice.[26] That would be hard to argue. The caricature displays a "preconceived opinion not based on actual experience"[27] casting Native Americans in a disparaging light. Moreover, in doing so, the caricature casts Native Americans – the "Indians" – as inferior. Finally, if I were to object that the caricature is not racialized, the bright red emphasis put on the color of the figure's skin would be hard to explain away.

Then imagine me insisting and shifting to colonialism. What is colonialist about a figure of Native Americans that could[28] be racist? I might say that I have no interest in maintaining colonization, and that, moreover, Native Americans have their own land on reservations – are full U.S. citizens, too. But if you pointed out that this state of affairs is actually unwanted by many indigenous nations here and that the U.S. government actively violates its own treaties with indigenous nations, things would become more complex for me.[29] If you could show that there are clear signs of ongoing domination – the clear and coercive repression of self-determination – then the caricature of the "Chief" would be overlaying a state of power-relations in which Native Americans remain colonized.[30] Worse, it would be making fun of Native Americans while serving up a racist stereotype.

It would be hard for me to dismiss the claim – and still make sense – that "Chief Wahoo" on a U.S. flag doesn't display a colonizing mentality.

However, the lore of being arbitrary is deep in the culture of the United States of America, at least as deep as the "Wild West."[31] We might not get very far considering what is really going on in the world and between each other. One of us – probably me, who has a harder argument to make and the risk of moral ignorance to bear if I don't make it – might become arbitrary and retreat behind property relations on my own land, protected by a law of liberty that is structured as freedom from interference.[32] Don't I have rights you have to consider? We might fail to be accountable with what we confront together, because my liberty would permit an arbitrary dismissal of your disagreement.

Autonomy, however, cannot do that. Here is the problem with arbitrary liberty: I'm not accountable to myself, I'm not accountable with the world, and my insistence on liberty over autonomy also involves a loss of community. The core of autonomy is sense – what makes sense to you – and sense is shared for us to make sense socially. From myself to others, back and forth between all of us through disagreement especially, sense roils us and between us. Autonomy follows sense where it goes.

Worlds are slippery between people, not uniform and fixed. Sense relates us to the world, potentially troubling any of our worlds. Each one of us may dislodge another's world by bringing new considerations to light about the world as we see it. If another has actually attended to the world, has tried truly to figure out what makes sense to them and why, then the line connecting them to their world – so to speak – can dislodge the line connecting myself to my world by raising a question about the world. When the world's in question – not just mine and not just yours – each of our worlds must be restored through consideration of what has arisen anew out of the possibly unknown.

When we do this together and consider what we each have thrown into the mix, we move toward articulating *our* world. Since autonomy cannot be world-avoidant (because sense cannot), autonomy, at the heart a practice of sense-making, cannot avoid the disagreement of others who dislodge any one person's complacency. Autonomy, gripping sense by the tail, thus is dragged and in turn drags us out into open community by the power of disagreement. My arbitrariness would undermine this.

Consider, then, this irony: Insofar as we seek to remain autonomous, we are each dragged into community by the power of disagreement.[33] If I believe that the world is flat, you show me that the world curves away from us, and I ignore you or do not have reason why the curve is illusory and the flatness is not, then I no longer make sense. What I am claiming does not respond to the world or to others.

This is what it means to say that sense is, at bottom, common – there for us all. To speak of common sense is not to imply that it is fixed, but that it has to stick between us and the world. If I want to remain autonomous, I need to square up to what you said about things, and this means also being accountable to you. A triangle, so to speak, emerges between each of us and the world, and we open an active and bare form of community around the process so delineated.

Community and common sense are intertwined and conflictual processes, not static things. Certainly, there can be things between us that cannot be resolved by attending to the world we share. There can be intractable disagreement. But I am skeptical of claims that this point arises early on in a disagreement, since it's often the case that someone has simply decided not to continue considering things, rather than realizing that there is no way to figure out anything further. Arbitrariness is a good explanation for why the processes often stop. Even so, when differences are irreconcilable, the plain truth that we have no clear answer here still shows sense we share. We share the realization that we cannot be sure of what makes sense here. That is the shared sense. At the heart of autonomy, there's much to go on when we stop making sense: active, open community as a conflictual and dynamic process.[34]

A week later, Shaker Heights, Ohio

Your objection breaks open common sense. If I am to keep making sense, I must consider what is truly common between us in the world that we can make sense of, considering it and each other's objections. As I suggested last week with the expression, "active and open community," this "we" – the "we" of a relationship – is the entrance of community as an intentional and living actuality, not as a fantasy or a presumption.[35] The conflictual and dynamic "we" can be trusted, because in autonomous disagreement, there is the commitment to work through it. The "we" is formed by the process of disagreeing. Through our commitment to stick together in disagreement and to respect each other as having our own considerations, we develop a sense of the common as something constructed from what is there without repressing our concerns. The world reveals itself sincerely through our disagreement as best as we can fathom it.[36] At the same time, by committing to make sense of how things have stopped making sense between us, we develop a sense of each other as trustworthy in finding what is common. Slowly, we find our world.

This is what I mean by claiming that a basic form of community begins here. Seen in the light of autonomous disagreement, confidence in community shouldn't really begin until we are assured that people are accountable to and in disagreement. Autonomous disagreement is the key to community, not the sign of its absence.

Now the notion of community in disagreement assumes a prior moral accountability to each other, and this must be brought out to understand why trustworthiness emerges through autonomous disagreement.[37] I think of this accountability as a basic moral relation.

A basic moral relation between people is involved in being a person at all. To discuss being a person is to assume the interpersonal.[38] The personal and the interpersonal go hand in hand. How could I be a person if I ignored other people? That would mean ignoring what I expect from other people as a person: recognition of my own personal capacity, experience, limits, and world.[39] That recognition is recognition of my being a person by other people. If I expect it, I must acknowledge the expectation in others, since they are people, too.

That much seems straightforward, but we can sharpen the point. Think about what goes into "being oneself," something that is part of being a person.[40] As I've already claimed, being oneself is an enacted, reflexive operation in which I am accountable for what I feel, believe, or intend.[41] *This is what I think. I want to do this. I am feeling this way now.* "The self" isn't a set of memories, a body, or a feeling.[42] It is an operation, and as such it shows up in being enacted. What is called "the self" is the misleading nominalization of the Latin *se*, a reflexive function in grammar that accompanies a verb. Similarly, as enacted, being oneself[43] is a reflexive operation of accountability in which I show up in the world alongside others and can be counted on regarding the implications of my thinking, intentions, even emotional awareness.

If X implies Y, and I believe X, then I will hold myself to believe Y once I see the implication. And if I am angry and I realize that I am angry, I will not tell myself that I am not; I will acknowledge my emotions. Finally, if I intend B, I won't instead claim that I intended C. In this way, being oneself is an elementary operation of accountability before myself and before others in which I reflexively commit to what it is that I assert, intend, experience, and so on.

Being myself, *I* claim a memory as decisive or important to me, commit to responsibility for *my* body, or acknowledge or admit my feelings. But such commitment would make no sense if I weren't seeing myself *as another* and asking, what do I stand for? Where am I? What is it to me?[44] So, being myself, I'm already in relation to others through the way I enact my commitment.[45] I seek integrity before myself, even, as before others.

Two of the assumptions of being oneself is that one is not another and that there *are* others.[46] This is called "differentiation."[47] When I'm myself, I differentiate myself from others, thereby being accountable to them as different – as having their own selves (or so I would expect). The process goes both ways. As I am capable of being myself, I'm also capable of accepting that others are, potentially, their own selves and should be treated as such unless evidence proves otherwise.[48]

Self-accountability and accountability to others thus go hand in hand in differentiation. Insofar as others are themselves and not me, I cannot think for them, act for them, or feel for them. That's their work, their life. To recognize this is to respect them, whether I like them, esteem them, or not.[49] I cannot be them. Their work is to be themselves.

From this ontological fact arises a moral relation that is already implicit in it. The relation is one of fundamental accountability to others as the selves that they are or will have to be if they are to be themselves. "The self" is structured morally in this elementary, relational way.[50] Being a self is differentiated from being another and implies both holding other selves accountable and being held accountable by others for being oneself. This is all out of respect for what it is for people to be themselves.

When you disagree with me, a moral relation of accountability is already with us. In its very form, disagreement involves respect. But notice what autonomous disagreement is and what it isn't. Disagreement elicits others as themselves. It's not something that overrides, undermines, denies, ignores, or otherwise throws off "the self" in others or in oneself. That would not be disagreement – it would be manipulation, effacement, suppression, disruption or some such other low-grade violence or disrespect. Someone who says that they are "disagreeing" but who doesn't support differentiation is possibly mistaken, deluded, or deceptive. At the least, they aren't clear with themselves. So they are not clear with you or others, either.[51]

Still, the community in disagreement is perhaps strange to some. Disagreement is differentiation expressed. It is moral. Orbiting around a difference, disagreement is respect in action for that difference. When we disagree, we don't see the sense in each other's worldviews at this point we share in the world. At the limit, we may not *see* that we share a world, although by disagreeing we already are disposed to share some (possibly new) world by the form of our commitment to make sense of things together. Not seeing the sense in each other's worldviews – in what we said, did, believed, showed, felt, made, imagined, etc. – we contest that sense. Contesting it, we challenge each other to consider what makes sense, starting by hearing out – attending to – each other's considerations on the matter. For the reasons I've been explaining, I claim that this is fundamentally respectful.

Think about what is implicit here:

You can make sense, and I can make sense. We are both intelligent.[52]
We are actually going to figure out what makes sense together, like it or not.
    We're going to articulate some common sense together.[53]
You have a perspective, and I have a perspective. We're different.[54]
You can count on me, as long as I am faithful to the disagreement, to be in
    community with you.

The relationships we form in autonomous disagreement are never one-sided. They involve multiple sources of consideration. The first source I've gone into; it arises from being oneself. We might imagine the "auto" in "autonomy" as a Greek version – *autos* – of what we find in the operation of the Latin *se*. It indicates a relation to oneself. The "nomy" in "autonomy" refers to the Greek word for norm – especially as custom, law, morals – *nomos*. It is the way one should do things. The way we might put these parts together is to say, *autonomous is the norm that makes sense to oneself.* You can commit to it; it is legitimate in your eyes – according to yourself. But for it to do that, you must implicitly accept differentiation – the respect of the selves of others, *their* autonomy, and vice-versa. Autonomous disagreement thus implies relationships in differentiation, accountability to the different selves of others as one commits to what makes sense to oneself and vice-versa.

The second source of consideration arises from elsewhere. Autonomous disagreement demands accountability with differentiations of sense. When people disagree, they do this in a question, implicitly or explicitly: What really does make sense, and why? This other person has a view themselves and to be autonomous myself, I must respect that they can have that view, that they can think, can feel and acknowledge their feelings, and that they can commit to intentions. Doing so, I have to follow out their sense by truly considering it. Our disagreement is thus an open question about what does make sense and why – to both of us, too, not just to me. The source of consideration is now the world between us, unless we arbitrarily avoid it.

But beliefs are delusional unless they square up with what they concern. So, too, with thinking. The power of thinking – the "I can think" – appears only when something calls for articulation within my mind and I try to understand it.[55] Similarly, intentions are seldom better than wishes unless they square up with what is in the world. So, too, with acting. The power of acting – the "I can act" – appears only when there is a world and a thing that is wanted to be done that I then try to navigate and do. Emotions, too, are little but sentimentality unless they respond to life and convey the emotional intelligence of what is happening. So, finally, with feeling. The power of feeling – the "I can feel" – appears only when there is something that affects me in life, something that demands an emotional and intuitive intelligence. Thus being oneself doesn't simply imply differentiation from others. It implies the resistance of the world (and of all its life) against whim, delusion, or sentimentality.

Being ourselves draws on resistance.[56] On the world's mysterious basis, we try to make sense. The community of autonomous disagreement is in this way grounded in the world's mystery. About it, we can only wonder.[57] Since searching consideration of the world by more than one of us is at the heart of the process of sense-making and also of what is held in common, we cannot properly speaking be arbitrary. Rather, being ourselves involves searching for sense. This disagreement about the world we happen to share demands figuring out what makes sense of the world's mystery between us. The common is then raised to view as open to the world's mystery.

Disagreement – differentiating each of us from each other and holding us in consideration of the world – keeps us in common around our searching. Multiple points of considerability produce multiple points of accountability grounded in respect for each other and of the world's mystery. Autonomy is accordingly far from being the property of an arbitrary individual, as it seems to be in liberty, but is a process of relationship that each individual has with themselves, others, and with the world.[58] It is communal.

All this contradicts the commonsense definition of "autonomy." One of the most common meanings of "autonomy" is independence from interference or control.[59] But the more I think about it, the commonsense definition seems to bear the traces of liberalism and to be consistent with capitalism. In actually existing capitalism, people are in competition with each other, and this makes it so that people are on their own in their basic economic relationship with each other.[60] But that is liberty – freedom from interference – and it is consistent with arbitrary behavior so long as one does not harm another or damage their property.[61] Such a view of autonomy creates confusion on confusion. There can't be accountability to each other without a basic moral relation. Is part of articulating anthroponomy wresting autonomy free from liberty, license, and capitalism?[62]

Contrary to the common sense in my liberal, capitalist country, relationships of autonomous disagreement are multifaceted, multi-sourced, and constructive. Since I can think, you have gotten me to see my flag differently. I was not considering things about it that actually go against my beliefs about what is reasonable here in our community. Your resistance, pointing to the world and the life we share, led me to be myself more by thinking, that is, by holding myself accountable with what is in the world, what I believe, and with my commitments.

Oddly, this resistance brought us closer together – around the world and the life we share, at least here in this disagreement.[63] Native Americans have widely rejected the domination of the settler colonial history of which the United States of America is the current and ongoing manifestation. That is independent of us, and it is true. "Chief Wahoo" is prejudiced and displays Native Americans – by racial marks – in a disparaging and foolish light, effectively sub-humanizing them. That is clear to anyone who attends to the details there in the caricature. Putting ongoing colonialism around and underneath – as context to – the racist caricature surfaces how the caricature happens to extend colonialism as an ideology, making its history and ongoing actuality invisible, keeping Native Americans as less than equal and less than respect-worthy in an order that benefits from the ongoing colony of the United States of America over native land and its plenty. But is it the norm of our community to see people as subjects and politics as maintaining the colony? We have to figure this out together – and to be open to disagreement coming from outside us, shaking our world here in Shaker Heights, Ohio.

Two weeks later, late evening, Shaker Heights, Ohio

I now have part of an answer as to how I should engage in community politics. To be engaged in community, I should disagree, and I should be open to autonomous disagreement. If I live my life in disagreement – disagreeing and being open to disagreement – I will engage in constructing an active and open community. Unity is problematic, whereas disagreement is related. A disagreeing community is the only kind worth trusting.[64]

Trusting in a disagreeing community, how should I understand politics? "Politics" comes from the word for city or city-state, *polis*.[65] It is conventionally associated with questions and matters of governance, i.e., the governance of a city or city-state. Governing, in turn, comes from the Greek word to steer, *kūbernan*. The idea is that the political concerns the steering of the city-state. When we translate *polis* to *community*, the political points to how community directs itself or is directed, that is, to its norms and its relations, institutions, and implements of authority. "Community politics" then points to the way we govern our world. We do so through community. Our "steering" of our community arises through norms that make sense to us in common, surfacing authentically only when and as we work through disagreement grounded in moral relations between us and the mystery of the world.[66] The question of how I should engage in community politics is the question of how I should help steer my community with and through disagreement in relationship with others who also disagree.

Given what I've clarified about active community, its politics should involve conflictual self-governance, "we"-governance,[67] whether in a village, a city, or a region. This seems promising for anthroponomy. But how should I think about such conflictual "we"-governance? One way is through what happens to common sense in disagreement. Since with politics, the norms, rules, or regulations "regulating" us presume what "us" is in common, and since there is nothing authentically common before disagreement, "we"-governance would seem to call for disagreeing, collective consideration of what makes sense to each of us, held by accountability to each other and with the world's mystery.

Steering life together through norms that go through a process of becoming authentically common, community politics is dynamically uncertain and isn't presumable in advance. The possibility of disagreement grounds it as potentially authentic, and the actuality of disagreement realizes it. Paradoxically, the uncertainty develops trust due to disagreement's implicitly moral relationships holding through the contestation. We hold together through confusion. Politics is morally formed around it and opens up around the disunited states of community life.[68]

Trying to figure out how to construct norms to govern where we live and to steer our world, we disagree. I genuinely do not see the world the way you do, but this does not mean that I want to stir up a fight or be contrarian. It means I am stating what I hold to be true, on consideration, not what I quickly pipe out of my mouth because something about you, me, someone else, or the world irritates me. In this disagreement, I am open to your disagreement, too. I am not simply making space for my own. This disagreement is neither mine nor yours, but is a process that opens when our worlds are disagreeable. Sometimes it opens on my end, and sometimes it opens on yours. Always, it is between us. In disagreeing, moreover, we hold the contention. I'm not simply fracturing the common sense between us, objecting to your world. We hold that fracture, that objection, while we can consider the world, each other, and others, across space and time. What makes sense for governing our common life across time? Our disagreement is a temporally extended process of consideration, not an instant shock of division.

Community politics aims to be trustworthy politics. People who can hold disagreement and who then work through it are people with whom one can have community. A person who holds disagreement takes their time with it and allows it to be part of life. They give it space and air, so to speak. They see it as important and will protect it. They may even support it. When they work through contention, they don't just burn through it. Realizations emerge in this process, not finger pointing. The "hold" and the "work through" of community politics are serious. Community politics brings out the trustworthiness in politics by situating politics in community relationships.

We can see how substantive this conclusion is if we compare it to liberalism, which structures my community to its own detriment.[69] Remember that someone's liberal right doesn't imply their autonomy. Autonomy differs from "freedom from interference"[70] in finding arbitrary will objectionable in relation to others. This is different from a liberal view of will. Liberty is a protected space to do and think whatever and however the hell you want, provided that you don't infringe on someone's else's liberty or harm them.[71] The name for this is, precisely, *license*.[72] You can be fully arbitrary in your license (and that may even be protected by a license). You don't need to make sense or to have any reasons for what you are doing or how you are doing it, including what and how you think or relate. The problem is, then, how can we share a world?[73] We barely will. We won't trust each other or the processes by which we find and make sense of the world. We will count on our liberal rights and self-ownership instead.

There is a strange, solipsistic individualism in the willful arbitrariness of liberty. It isn't intentionally selfish, but it is self-indulgent and atomistic. Autonomy, however, is communal, relationship-based and responsible. Unlike liberty, autonomy involves a commitment to relate to the differentiation between people and to the people themselves. Autonomy also involves accountability with the world and the life we share, even in its mystery. In being self-respecting and self-accountable, autonomous people must also be other-respecting and other-accountable; they must be world-accountable. With autonomy, precisely, we are in the world together.[74]

People of liberty may forget this planet in their self-ownership provided that they are not directly harming others. But people of autonomy must be responsible with it. This makes autonomy political and planetary in a way that liberty simply cannot be. Far from being solipsistic or self-indulgent, people of autonomy must be grounded in the mystery of the world and be related communally to each other through the formative possibility of disagreement. They are primed to be planetary by the nature of their relatedness.

I want to linger on the importance of relationships. Community politics can't be a state of competition or a state of war among interests.[75] It's a process involving an intention. The intention is to find norms governing us that make sense to us together.[76] We try to construct the trustworthy basis of these norms through accountable and autonomous disagreement around how we can make sense of the world in common.[77] The trustworthiness of the norms we construct draws on and reinforces the primacy of moral relationships in the differentiated process of disagreeing. It is also reinforced by taking place around a mystery about which we must be accountable. Not one of us can make absolute sense of the world. The world's mystery, demanding that we make sense of it, is at the heart of politics between us.[78] The relationships in community politics are moral and epistemological, involving trustworthy values such as respect for persons, truthfulness, and even humility.

*

Some weeks ago, I went out on a run in a world of liberties and was dumbstruck by their inconsiderateness. There was "Chief Wahoo" on a flag. Seen from the point of view of one standing on the side-walk stunned by a racist and colonialist identification of a nation trumpeted for it "liberty," I have now found that there is something novel about autonomy as a guide to politics.[79] Through autonomy and its relationships, the political appeared as accountability to disagreement over the norms of a community we cannot yet call ours,[80] but which grounds us in the world's mystery.[81] This politics can be scaled to a household just as it can be scaled to a state – or to a planet, in principle. I am beginning to realize how autonomy helps clarify what anthroponomy can be.

Called "Kahyonhá:ke"[82] by some, "Cleveland" as I learned of this place, late in September

I wanted to know how I should engage in community politics, and now I have the beginning of an answer. In this age challenged by planetary scaled issues that appear so much to divide us or leave us isolated from each other, I should work to protect disagreement to be as open and as deep as it is around the norms that govern our lives together. This disagreement, contrary to common sense, is consistently, morally thoughtful and appreciative of the world's mystery. It opens us up to the world and to each other.

I can sense how I am in the neighborhood of anthroponomy now. Anthroponomy is my response to the "Anthropocene" – a word that buries violent and dominating histories such as colonialism by making all of humankind socially homogenous in its causal relation to our planetary situation. One thing good about protecting disagreement is that it helps bring into the open the history of colonialism buried by talk of the "Anthropocene." The disagreement surrounding the unworked through history of colonialism is everywhere around me when I listen for it. It is there on a plain run through my city. It points to an injustice that is old, deep, and persistent. Looking forward to our planetary future, disagreement around colonialism is crucial. Community politics depends on it, including the autonomy without which there is no form of anthroponomy. Sometimes the ways forward is the way back.

Now the main obstacle to anthroponomy we've inherited through colonialism is domination. Still existing colonialism in my nation covers over its own domination, as the flag showed symbolically. "Chief Wahoo" buried the truth. But in its very form, open disagreement is already resistance to domination. Built into it is sense-seeking disobedience and refusal of forced unity. In a world where there is domination, coercive inequality, and intimidation, to engage in a process against such things is morally and practically important. Only when there is freedom from tyranny can there possibly be a relationship of equality with another, let alone trust,[83] and any future world that is predicated off of domination is an unjust one. Disobedience to domination through disagreement emerging communally between people is a world where colonialism increasingly finds itself challenged by moral thoughtfulness.

Many of the major nations of the world are ongoing colonies, shorn free from their historical empire. This includes the United States of America, Canada, Australia, and New Zealand. The distribution of wealth and power across most of the globe also continues to raise questions about the relationship of equality between Europe and many other nations on other continents–Africa, South America, Central America, the "Middle East" and South Asia[84] – that were formerly colonized by

Europeans. The culture taught in many schools throughout the English-speaking world still reinforces colonialism, too. To challenge histories of domination in colonialism's legacy involves a releasing of names, cultural memory, and cultural orientation as well as a challenging of the order of nations around the planet. Its possible consequences are, quite precisely, world-shaking. What worlds should arise from these quakes?

I think, for example, of where I live, "Cleveland," as it is called in my settler colonial society. The legacies of colonialism are thick, overlaid, and violent in this city. Domination as a history is woven into the organization of the city, its infrastructure, administration, wealth allocation, effective rights and permissions, opportunities, official history, and imagery. There are many, traumatic strands to it. Two large, woven ropes are the ongoing dispossession, scattering and relocation of Native Americans from many different nations, here in Cleveland without an institutionalized center or land, and the deep presence of the life of descendants of slaves, configured and disfigured by the ongoing inequalities and injustices of areas in Cleveland that were formerly "red-lined."[85]

When I think of what it would be to protect disagreement about what makes sense in the world from within the many inequalities that structure and derive from the historical and present urban organization of Cleveland, it is as if I hear ghosts erupting from the Earth and rising into air. I do not have another way to explain in a manageable way the impression of the trauma, depression, anxiety, fear, rage, and indignation that are – legitimately – thick in the bones of this city. I say, "legitimately," because in no way is it obvious that this city has ever been equal, safe, and autonomous. The domination of colonialism still bears down like history's meteor storm on the Black bodies around this land: poverty, unemployment, derelict housing, homelessness, ill-funded education, few opportunities to develop agency or wealth, violence from Black on Black crime, drugs, alcoholism, high incarceration rates and what Dr. Martin Luther King Jr. called a "toxic" poison of low-self-esteem and worthlessness, a bone-deep anomie.[86]

Being in Cleveland, it's clear to me that if I'm to work toward anthroponomy in a city such as this one, colonialism's continuing presence obdurately stands in the way.[87] Capitalism and industrialization have shaped my city profoundly in concert with this legacy of colonialism. Colonialism's legacy explains to a large degree the forms, uses, locations, and permissions of capitalism and industrial production in this city, especially through racism.[88] Colonialism, sunk into oblivion by the eerie certainty of "Chief Wahoo" and its wider colonial ideology, fits the history of violence right here in the land where I live.[89]

To engage in community politics here in this haunted, still dominated land is to find disagreement from out of a kind of oblivion. I wonder if this is why my reflections have at times been hard and sluggish as if I were a sightless animal lacking the sense of smell. I want to look toward the future and do something about it, something constructive, but to do that I must look toward the past and do something deconstructive. I have to side with "five hundred years of resistance"[90] and make way for the autonomy of disagreement to unsettle the liberal arbitrariness of my city.

I responded to the "Anthropocene" with anthroponomy. Now I realize that anthroponomy begins with disagreement over the erasure of colonialism right *here*. Beginning here in my would-be community, I have to learn from something as twisted and deep as colonization, a particularly deep and pervasive world-scaled form of domination that will negate anthroponomy. The *via negativa* to anthroponomy disappears into colonization's oblivion.[91]

A day later, Shaker Heights, Ohio

Whether out of dreams or the night's stillness, I don't know, I woke in the early morning to the sound of leaves rolling on trees through gusts of wind. Wind was oblivion, the autumn's collect.[92]

[. . .]

How should I begin to relate to colonialism?

Notes

1 That is, "if we do not have a morally acceptable relation to our planetary situation."
2 Charles Larmore, *The Practices of the Self*, translated by Sharon Brown, Chicago: University of Chicago Press, 2010.
3 Or any similar thing claiming that we *should* or *must* follow or accept it.
4 Jean-Luc Nancy, *The Inoperative Community*, translated by Peter Connor, Lisa Garbus, Michael Holland, and Simona Sawhney, Minneapolis: University of Minnesota Press, 1991. But repression is indirect, too. It can be hard to know one is doing it; one becomes accustomed to blocked off parts of oneself, to voices one dismisses in one's mind. Perhaps, then, we need indirect ways to allow ourselves to open up, ways into our own oblivion.
5 Cf. Kyle Powy's Whyte on the "politics of consent" as opposed to the "politics of urgency"; "Disrupting Crisis, Unsettling Urgency: An Indigenous Criticism of Assumptions about Time in Environmental Advocacy," *Balsillie School of International Affairs*, March 11, 2019; also, the CENHS (Center for Energy & Environmental Research in the Human Sciences) podcast 166, Rice University, February 28, 2019. Whyte's notion of the politics of consent and the disruption of the politics of urgency counterbalances the apparent politics of urgency in works such as Geoff Mann and Joel Wainwright, *Climate Leviathan: A Political Theory for Our Planetary Future*, Brooklyn, NY: Verso, 2018 or even in Stephan Gardiner's *A Perfect Moral Storm: The Ethical Tragedy of Climate Change*, New York: Oxford University Press, 2012, which imports both capitalism and the colonial world system as unexamined normative assumptions.
6 Martin Luther King, Jr., *Where Do We Go from Here: Chaos or Community?* Boston: Beacon Press, 2010, p. 28. Dr. King is speaking of a "happy life," to be precise, and is doing so in the context of white backlash against civil rights. The full passage reads:

> The answer [to the dilemma of how blacks can work with whites when there is persistent inequality] was only to be found in persistent trying, perpetual experimentation, persevering togetherness. / Like life, racial understanding is not something that we find but something that we must create. . . . A productive and happy life is not something that you find; it is something that you make. And so the ability of Negroes and whites to work together, to understand each other, will not be found ready-made; it must be created by the fact of contact.

7 Fractal forms reiterate the same pattern within themselves. The *Oxford American Dictionary* (Apple Inc., version 2.2.2., 2005–2017) states:

> **fractal** | ˈfraktəl | Mathematics *noun*
> a curve or geometric figure, each part of which has the same statistical character as the whole. Fractals are useful in modeling structures (such as eroded coastlines or snowflakes) in which similar patterns recur at progressively smaller scales, and in describing partly random or chaotic phenomena such as crystal growth, fluid turbulence, and galaxy formation.
> *adjective* relating to or of the nature of a fractal or fractals: *fractal geometry.*
> ORIGIN 1970s: from French, from Latin fract- 'broken', from the verb *frangere.*

8 Cf. "Cleveland Indians Name and Logo Controversy," *Wikipedia*, 2018.
9 Mike Sauter, "The 16 Most Segregated Cities in America," *24/7 Wall Street*, July 21, 2017; Cleveland is there at #5. In 2015, it was # 1.
10 According to the U.S. Census Bureau as of July 1, 2017, Shaker Heights includes 34.3 percent African Americans, 54.7 percent whites, and 0.2 percent Native Americans or Alaskans; "Quick Facts: Shaker Heights, Ohio," *United States Census Bureau.* In addition, some of the areas of Shaker Heights have higher proportions of African-Americans and lesser proportions of whites.
11 I put "Americans" in scare-quotes, because of the moral ambivalence of the term in the context of colonialism. The term "America" is a colonialist cartographical name covering over the indigenous names of lands. Who feels comfortable with its name – and who feels uncomfortable?
12 "Chief Wahoo as a Racial Caricature" in "Cleveland Indians Name and Logo Controversy," *Wikipedia*, 2018.
13 Cf. Kristie Dotson, "Abolishing Jane Crow," *Social Epistemology Review and Reply Collective 7*, n. 7, 2018, pp. 1–8, specifically on "being disappeared." See also Kyle Powys Whyte, "Settler Colonialism, Ecology, & Environmental Injustice," *Environment & Society*, 9, 2018, pp. 129–144.
14 Note that liberty is not the same as autonomy.
15 The idea is to articulate a *prima facie* sense of autonomy that stands up to ordinary life, even if we have not yet gone into the many subtleties of the concept of autonomy once it is philosophically analyzed and historically delineated as it is in texts as various as Christine Korsgaard's *The Sources of Normativity*, New York: Cambridge University Press, 1996; J.P. Schneewind's *The Invention of Autonomy: A History of Modern Moral Philosophy*, New York: Cambridge University Press, 1998, or Axel Honneth's *Freedom's Right*, New York: Columbia University Press, 2015.

For further background reading, see John Christman, "Autonomy in Moral and Political Philosophy," *The Stanford Encyclopedia of Philosophy*, spring 2018, accessed July 24, 2018. See, for instance, the *leeway* a plain understanding of autonomy might enjoy in this remark from Christman (in "1.1. Basic Distinctions"):

> In addition, we must keep separate the idea of basic autonomy, the minimal status of being responsible, independent, and able to speak for oneself, from ideal autonomy, an achievement that serves as a goal to which we might aspire and according to which a person is maximally authentic and free of manipulative, self-distorting influences. Any plausible conceptualization of basic autonomy must, among other things, imply that most adults who are not suffering from debilitating pathologies or are under oppressive and constricting conditions count as autonomous. Autonomy as an ideal, on the other hand, may well be enjoyed by very few if any individuals, for it functions as a goal to be attained.

Christman (in "I. The concept of autonomy") associates autonomy with authenticity, the "true self":

Put most simply, to be autonomous is to be one's own person, to be directed by considerations, desires, conditions, and characteristics that are not simply imposed externally
upon one, but are part of what can somehow be considered one's authentic self.

However, following Charles Larmore, all authenticity is, I think, is *acting as a self*, that
is, being responsible to the implications of what one believes is true, what one intends
to do as good, and what one feels, especially in relationship. That's subtler. I'll employ
Larmore's work further as this chapter progresses. See Larmore, 2010.

16 See *Five Hundred Years of Resistance and Dignity* ("Change the name! Change the
logo!"), and "Chief Wahoo as a Racial Caricature." There have been protests every
year for decades at Cleveland Indians games by *Five Hundred Years*.

17 Cf. my "Reconsidering the Aesthetics of Protest," *Hyperallergic*, December 7, 2016.

18 Larmore, 2010.

19 Judith Butler, *Giving an Account of Oneself*, New York: Fordham University Press, 2005.

20 Cf. Charles Larmore, *The Morals of Modernity*, New York: Cambridge University Press, 1998 and the theme of disagreement as the expression of increasingly
autonomous – or at least politically liberal – people. Disagreement is not a result of
egotism; it is the result of being oneself, i.e., having to make sense to oneself of things
from one's perspective and not simply conforming to what one is presented with.

21 Nancy, 1991.

22 John Locke, *Second Treatise of Government*, Indianapolis: Hackett Publishing, 1980,
chapter 2, section 6, however, is clear on the face of it that liberty does not imply
license. What Locke means by this is that liberty to use one's property as one will is
not unconditional. The use of it cannot harm another or destroy nature unless it be for a
"nobler" purpose. And yet the "lower" animals are "made" to be "used by us." Such a
distinction between liberty and license doesn't seem to imply that one must be morally
accountable to others in matters that do not involve harm or destruction to their property or to their bodies, which the racist flag does not seem to do. It also narrows license
to the destruction of one's own property for "lower" purposes. These distinctions in
Locke are to be separated, too, from a roughly Lockean society in which individual
liberty legally permits racist license when direct violence to the bodies or property of
others is not part of it, again as is the case with the flag.

23 Isaiah Berlin, "Two Concepts of Liberty," in *Four Essays on Liberty*, New York:
Oxford University Press, 1969, pp. 118–172.

24 If you disagree with my view of autonomy, advance what you take to be a better view of
liberty, explain why you do, and expect me to hear out your reasoning, isn't it interesting that, in so disagreeing, you actually confirm autonomy's useful primacy? One might
describe my view as running through the landscape of *democratic epistemology and
accountability* unlike Russell Hardin's (see Russell Hardin, "Democratic Epistemology
and Accountability," *Social Philosophy and Policy*, v. 17, n. 1, 2000, pp. 110–112).
Hardin focuses on the scale and specialization of knowledge in nation state governance. I follow out ground level democratic epistemology between people in community
improving itself through accountability in disagreement. My thanks to Michael Ignati-
eff, discussion, Central European University, Budapest, Hungary, January 2019.

25 Cf. my "This Conversation Never Happened," *Tikkun*, March 26, 2018, and "How to
Do Things Without Words: Silence as the Power of Accountability," *Public Seminar*,
June 28, 2018.

26 The *Oxford American Dictionary* (2005–2017) states that something being "racist"
means that it shows "prejudice" against "people of other races" or suggests "superiority" of one racial group over another.

27 "Prejudice," *Oxford American Dictionary*, 2005–2017.

28 To remain reasonable I would have to accept the first part of the objection.

29 To begin with, the Doctrine of Discovery, which undergirds the cornerstone U.S.
Supreme Court ruling around Federal and Native American land, *Johnson v. M'Intosh*,

21 U.S. (8 Wheat) 543, (1823), is rejected by many Native Americans. It presumes that the European powers that colonized "America" had a right to own the land they "discovered," thereby annulling the claims of prior belonging with the land of Native American nations. When the U.S. formed after rejecting English rule, the same "rights" were conveyed to the new nation as a transference of the prior colonial "precedent." For one overview, see Walter Echo Hawk (Pawnee), "Johnson v. M'Intosh & the Doctrine of Discovery in the United States–Impacts upon Federal Indian Law; and the Future of the Doctrine under the United Nations Declaration on the Rights of Indigenous Peoples," Shuswap Nation Tribal Council and Thompson Rivers University, September 20, 2012.

30 And you could, linking Canada to the United States so that all of North America is implicated, you could point to the Dakota Access Pipeline (DAPL) and the Tar Sands in Canada. See Faith Spotted Eagle, "Traditional Leadership from Mother Earth: Standing Rock and the *Mni Wiconi* Gathering – an Evening with Faith Spotted Eagle (*Ihonktonwan Dakota*)," Case Western Reserve University, October 18, 2018 and Glen Coulthard, *Red Skin, White Masks: Rejecting the Colonial Politics of Recognition*, Minneapolis: University of Minnesota Press, 2014.

31 See my "How to Do Things Without Words," 2018. Note the historical relation between the "Wild West" and colonization, removal, relocation and the mass killing, often intended genocide, of Native Americans.

32 Here the legacy of Locke, 1980, appears again, this time with respect to private property's relationship to liberty and liberty's relationship, if not to what Locke called "license," then to arbitrary relations to others in any matter that does not directly involve harm to them or destruction of their property.

33 If, for some who misunderstand it, "autonomy" means to live on one's own, then it would be ironic that autonomy drags one into community whenever there is disagreement over the sense of the world.

34 There is, of course, more to community than working through disagreement together. Disagreement isn't sufficient for community in a robust sense. But it does appear to be necessary.

35 Cf. Nancy, 1991 and the notion of the common as an ever-worked-out relationship among a plurality of people, not a presumed *ethos* that projects a unity of people around the same views. See also my "Democracy as Relationship," *e-Flux Conversations*, April 30, 2017.

36 "The world reveals itself sincerely" is a term of art for the quality of the world *as found by us* when we are forthright and without reservation about our concerns and doubts, open to our disagreement and that of anyone else who arrives in disagreement.

37 Obviously, the countervailing position here would hold that disagreement generates distrust. One of the indirect aims of my argument is to disagree, respectfully, with such a viewpoint. Disagreement understood within autonomous practices and relationships is, far from a breeding ground of distrust, an origin and cultivator of trust.

38 Cf. Stephen Darwall, *The Second Person Standpoint: Morality, Respect, and Accountability*, Cambridge, MA: Harvard University Press, 2006.

39 Cf. Axel Honneth, *The Struggle for Recognition: The Moral Grammar of Social Conflicts*, translated by Joel Anderson, Cambridge, MA: MIT Press, 1996.

40 Cf., Ralph Ellison, *Invisible Man*, New York: Modern Library Edition, 1994, where failing to be recognized as a person in a society struggling with colonialism's ongoing effects introduces challenges to being oneself, but where being oneself maintains one's *need* to be seen as a person and thus to be a person with others.

41 Larmore, 2010. See also my "How to Do Things Without Words," 2018.

42 That is, at best, "me," direct object. It becomes part of myself as a reflexive function only when I evaluate myself to see if "me" is consistent with what I believe, intend, etc.

43 "Being oneself" has the advantage of avoiding the misleading nominalization.

44 Cf. Paul Ricœur, *Oneself as Another*, translated by Kathleen Blamey, Chicago: University of Chicago Press, 1992.

45 Committing to oneself in relation to others is the opposite of self-possession, the political philosophy at the founding of the United States of America. In the tradition of possessive individualism, people "possess" themselves and have to "enter into a contract" to be related (and so need *not* be related if they do not want the contract). But one needs to possess oneself only when the primary relations with others have *broken down*. Self-possession assumes and internalizes a prior threat from others. Cf. C.B. Macpherson, *The Political Theory of Possessive Individualism: Hobbes to Locke*, New York: Oxford University Press, 2011; and my *The Wind ~ An Unruly Living*, Brooklyn, NY: Punctum Books, 2018.

46 Seeing oneself *as* another is not the same as *being* another.

47 Cf. Murray Bowen, "On the Differentiation of Self (1972)," in *Family Therapy in Clinical Practice*, Lanham, MD: Rowman & Littlefield, 2004, pp. 467–528; see also "Toward the Differentiation of Self in One's Family of Origin (1974)," pp. 529–548 and "The Use of Family Theory in Clinical Practice (1966)," pp. 147–182. See also "Differentiation of Self" under the entry, "Murray Bowen," *Wikipedia*, 2018. I am using the term differently than Bowen, of course, since I am not focused on differentiation in the family system, where I believe my use of differentiation is probably rooted. My use is more general and also less rich in capacities than Bowen's. You might think of my use as a *presupposition* of Bowen's and as a *generalized* achievement of work on differentiation in his sense through therapy or other processes of growing up.

48 It is entirely possible for people not to have a stable sense of self, that is, a stable capacity to be accountable for their intentions, beliefs, and feelings, for instance.

49 The form of respect is so basic as to not be anything more than what it formally acknowledges. It isn't admiration. It is consideration of the kind of beings that people are (cf. Darwall, 2006).

50 This argument for basic moral relation in the structure of being oneself is different than Emmanuel Levinas' argument for the "priority of the ethical" *over* the "I." Levinas insisted on the *loss* of the self, its "original traumatization" by the other. As Gayatri Spivak pointed out, Levinas's way of internalizing violence is problematic. In my terms, it distorts plain relations of accountability to oneself as to others. Cf. Emmanuel Levinas, *Otherwise Than Being or Beyond Essence*, translated by Alphonso Lingis, Pittsburgh: Duquesne University Press, 1998; Gayatri Chakravorty Spivak, "The Politics of Translation," in *Outside in the Teaching Machine*, New York: Routledge, 2009, pp. 200–225.

51 "Clear" here means being forthright and accountable.

52 Jacques Rancière called this, "the equality of intelligence." Jacques Rancière, *The Ignorant Schoolmaster: Five Lessons on Intellectual Emancipation*, translated by Kristin Ross, Palo Alto: Stanford University Press, 1991.

53 Articulation substitutes for the "co-construction" of understanding. Rather than make things up, we reveal the world together dialogically. Cf. Daniel R Scheinfeld, Karen M. Haigh, and Sandra J.P. Scheinfeld, *We Are All Explorers: Learning and Teaching With Reggio Principles in Urban Settings*, New York: Teachers College Press, 2008.

54 We have a perspective *on* something. I think that something makes sense in a certain way, and you think it doesn't or makes sense in a different way. We have to get to the thing between us to disagree effectively.

55 See the *locus classicus* of this point in Immanuel Kant, *Critique of Pure Reason*, translated by Norman Kemp-Smith and Howard Caygill, New York: Palgrave Macmillan, 2007. Kant does not think that we can grasp what is independent of the mind, only that the experiences of error and of learning make us aware that our minds grow in articulation outside of which we both cannot think yet cannot delude ourselves that what we think is without further potential for articulation.

56 Cf. Hegel's understanding of self-consciousness emerging from "negation" by objects in the most general sense. G.W.F. Hegel, *The Phenomenology of Spirit*, translated by A.V. Miller, New York: Oxford University Press, 1977.

57 See my essay, "Lostness," *Between Us: After Martha C. Nussbaum's Politics of Wonder*, chapter 1, New York: Bloomsbury, forthcoming.
58 Cf. Honneth, 2015, 1996.
59 "Autonomy," *Oxford American Dictionary*, Apple Inc., 2005–2017.
60 See my *The Wind*, 2018, esp. regarding self-ownership.
61 Again the province of Locke.
62 Would this be a kind of decolonial work?
63 Holding disagreement, holding each other in respect and holding things in common are linked. The idea here is that disagreement is a holding pattern for community, and consideration is a holding pattern for what is common. For more on what I have called "holding patterns" in the form that reasons of "common" humanity take, see my *Conscience and Humanity*, dissertation submitted to the University of Chicago Department of Philosophy, MI: UMI, 2002, IV.1.1.
64 That need not mean a fractious community. In communities structured by injustice, disagreement when it erupts is likely to be fractious. But communities are more or less just. Just ones might express disagreement in different ways, some of which may be even wonderful.
65 *Oxford American Dictionary*, Apple Inc., 2005–2017, "political," "politics," "govern."
66 See Misty Morrison, *Oblivion*, Pittsburgh, PA: Pittsburgh Center for the Arts, 2017. Morrison's use of the word and her emphasis on multi-perspectival disagreement even *within* oneself influences my use of the word here. She draws on Marc Augé's *Oblivion*, translated by Marjolijn de Jager, Minneapolis: University of Minnesota Press, 2004. Augé is strongly influenced by Nietzsche's "The Use and Abuse of History for Life" (1874).
67 Cf. Jean-Jacques Rousseau, *The Social Contract*, translated by Maurice Cranston, New York: Penguin Classics, 1968
68 What would be the "Disunited States of America?"
69 Its "self-ownership" is its *own* detriment.
70 See Berlin, 1969; see also for a wonderfully helpful contrast, Philip Pettit, *Republicanism: A Theory of Freedom and Government*, New York: Oxford University Press, 1997.
71 John Stuart Mill, *On Liberty*, Indianapolis: Hackett Publishing Company, 1978, Introduction.
72 Not license as defined by Locke, 1980, chapter 2, section 6; but license as enabled by a Lockean inheritance that severed liberty from relationships beyond non-interference, non-harm, and protection of property.
73 Any democracy I exercise then will be oddly without "us."
74 Cf. Honneth, 2015; my "Democracy as Relationship," 2017; Matthew Hodgetts, *We Are All in This Together: Addressing Climate Change Through Our Global Ethos*, dissertation submitted to Brown University Department of Political Science, 2017.
75 This neo-liberal, Hobbesian formulation is explained well in Wendy Brown, *Undoing the Demos: Neoliberalism's Stealth Revolution*, Cambridge, MA: Zone Books, 2015.
76 See my "Democracy as Relationship," 2017.
77 Cf. Steven Vogel, *Thinking Like a Mall: Environmental Philosophy After the End of Nature*, Cambridge, MA: MIT Press, 2015 on social construction as a process of building the world together through deliberative practices. My emphasis, however, is on disagreement – a way to understand antagonism within the search for community politics. Cf. William Connelly, *Identity/Difference: Democratic Negotiation of Political Paradox*, Minneapolis: University of Minnesota Press, 2002. One thing important in my view is the presence of the world. There is a triad in the space of contestation, that is, of disagreement: self – world – other. See also my use of Paolo Freire and Karl Jasper's limit situations in "The Neoliberal Radicals," *e-flux Conversations*, February 1, 2017.
78 Bruce Kafer, "A Climate Change on Climate Change," Hathaway Brown, Shaker Heights, Ohio, January 11, 2019 on the Lakota creation story centered on the mystery at the center of the world as seen by finite humans.

79 Cf. Alain Badiou, "The Neolithic, Capitalism, and Communism," *Verso Blog*, July 30, 2018; "The problem is not technology or nature. The problem is how to organise societies at a global scale."

80 Cf. Jacques Rancière, *Disagreement: Politics and Philosophy*, translated by Julie Rose, Minneapolis: University of Minnesota Press, 2004.

81 *Its* sincerity, so to speak.

82 *Kahyonhá:ke*, according to M. Carmen Lane. Susan Dominguez writes, " 'Crooked River' (the Cleveland area) was used by some of the Six Nations in their individual languages." Email correspondence, August 22, 2018.

83 Cf. Pettit, 1997.

84 It is worth noting that the geography of these lands is indicated in colonialist terms – from the notion of the "East" to the name, "America." Names themselves betray the point – they are the results of domination; and what would they become if there were made sense of again through protected disagreement?

85 That is, peopled by racist real estate practices that have kept descendants of slaves in abject poverty and suffering and which have created a settled impression of intergenerational poverty and self-destruction in many of Cleveland's and East Cleveland's Black neighborhoods. Some of the details of these claims can be found in the student-research project on anthroponomy in Cleveland, "V2V: From Violence to Violation," Case Western Reserve University, Spring Semester, 2018.

86 King, 2010, chapter 5, "Where do we go from here?"

87 See my "Decolonialism and Democracy: On the Most Painful Challenges to Anthroponomy," *Inhabiting the Anthropocene*, July 27, 2016. Colonialism, capitalism, and industrialization are bound together; yet each have a different logic that historically draw on each other at different points in the last half millennium of what was initially European history. Colonialism leads the way in my situation because of the clear presence of ongoing domination in its ongoing dynamics since the legal yet half-hearted abolition of slavery, and because its relation to the land in still settler colonial society permits the resource extraction used to cause an oppressive situation for future generations.

88 "V2V," 2018.

89 Cf. Kristie Dotson, "Tracking Epistemic Violence, Tracking Practices of Silencing," *Hypatia*, v. 26, n. 2, 2011, pp. 236–257 and with Marita Gilbert, "Curious Disappearances: Affectability Imbalances and Process-Based Invisibility," *Hypatia*, v. 29, n. 4, 2014, pp. 873–888.

 If the root of the past is in the future, it may be in the way the "Anthropocene" – that future word – erases the social processes that powered dominant nations and privileged systems within them to steer our world into a future that many of us fear for our descendants. There it is, the *past,* you might say, revealing a world system's inequalities of the far, far *future* where generations-to-come have to bear the burden of a reckless order. In this way, one root of the past – colonialism – can be seen in the future called the "Anthropocene." It is seen in the erasure of accountability for domination, right there in the futuristic name, not unlike the "Chief Wahoo" logo. The "Anthropocene" is a futuristic word that naturalizes the problems that created what it is about. We go to the "Indians" game and forget the poverty of our brothers.

90 *Five Hundred Years of Resistance and Dignity.*

91 The *via negativa*, way of negation, is an argumentative method of determining what something could be by marking what it is not.

92 The "collect" is a name for a call to prayer. Hearing "collective" with the "collect" in it, something collective is a putting-into-motion of the collect. Although anthroponomy is not a religious idea, the analogy is provocative. Perhaps disagreement is the collect of anthroponomy. After all, if there ever were an authentic collective, it would be held in a space where people *collect* in disagreement and protect its working-through. "Collect," *Oxford American Dictionary*, Apple Inc., 2005–2017.

Figure 3.1 Over Eastern Ohio, not far from Lake Erie, September 2018

3　How should I understand my responsibility and show it?

Shaker Heights, Ohio, autumn, 2018: a letter to my mother in Belle Valley, Ohio, 1939, filed in an old, metal filing cabinet in the basement

File tab: "Colonial families"

Dear Esther,

I know that the words I am using are not your words from 1939. In 1939, you didn't have words. You had just been born. I don't know our family's words from then. Many of them were in Slovak, a language I do not speak. Our Slovak disappeared as we were absorbed in this "New World." Still, I am imagining you, today, in an overtone of time.

What do I mean by an "overtone" of time? I can tell you how I began to think about it. I was trying to learn from an indigenous tradition as I thought about how I should relate to the history and ongoing presence of colonialism and the widespread problems of actually existing capitalism here in Cleveland. I was thinking about these things in relation to our planetary situation. As we often discussed, things are not right on this planet with the global economy and with nations such as ours. The dominant social processes of the world are putting at great risk the future generations of humankind, foremost those who will be most vulnerable due to powerlessness.[1] The world's dominant social processes are also throwing to the void the great mass of life, risking a mass extinction, which will decimate the order of life on Earth for millions of years – far beyond reasonable life expectancy for our species, too.[2] All of this is so appalling that despair would set in if it weren't that moral urgency demands a plain and relatable response. I want to figure my moral responsibility within these social processes.

One of the first things I want to throw off is the misdirection I see around us when our current planetary situation is characterized in mainstream media. It's increasingly characterized as the "Anthropocene" – the geological age in which humans shape, even drive, the Earth's biochemical, ecological, and geological reality in major ways. What bothers me about this term is that it obfuscates the social processes that are causing our situation,[3] specifically, a bundle of processes begun in early modern times, when colonialism lay the system for capitalism's expansion, corporations were begun that managed the colonies,[4] and industrialism – based on extracting fossil fuels and other resources from the land – gave rise to ongoing cycles of imperialism on a global scale. Human nature is *not* responsible for our planetary situation. Humans *as such* are not destroying the extant order of life on this planet. Rather, specific social pathologies are and the people that consciously support them after knowing what they do.[5] I want to draw attention to what is beyond the obfuscation of the "Anthropocene" in order to understand my responsibility accurately and to see the problem for what it is.

My way of carrying out this task is to focus on a notion we discussed more than once, *anthroponomy*. I think that instead of the "Anthropocene," we should be thinking about anthroponomy. Anthroponomy is the coordination of humankind across space and over time to protect people and their moral relations from domination. Since moral relations include the more than human world of other forms of life in varying ways and degrees,[6] anthroponomy amounts to an open-ended process of holding humankind accountable for its wantonness with the world of life and for its dominating social processes. Anthroponomy is a coordinated and specific orientation of our autonomy around the planet and over many generations of time.

Given the need to think about a time scale that is more adequate to anthroponomy, I began to think about "spiral time." "Spiral time" is the name Potawatomi scholar Kyle Powys Whyte gives to a way of viewing time in which ancestors and descendants are contemporaneous in the sense of time.[7] I was interested in this temporal understanding, its practice and the relationships opened up by it. It seems to throw off our colonial order of time, fixated by capitalism on the very near term, throwing away the past with each new investment and product.[8] Spiral time seems to be the kind of time that is readily anthroponomous.

Here is what I learned: To live in this world with the impression of my ancestors and my descendants living in the same place I am, only at different times. They are with me, and I am with them. There is a relationship we have by being in the same *place*, joined by the land in which they lived, I live, and my descendants will live. This place in which I live is the point intersected by the spiral. It is multi-layered and can include a place in life, a situation, even an emotional point. I am about to have a child now, just as Grandpa Bendik did with you. But this existential place still depends on the land that has supported us and which is a literal *placeholder* for the lineages of life constituting and inhabiting it. Ultimately, this place in which I live is, most simply in my language, the Earth. Being joined in place by spiral time layers existential connection over the land that makes that connection possible.

Spiral time lands relationships in place. When I drive down to Belle Valley or Pleasant City past Cambridge, Ohio, I feel our family more strongly. The land fits – or, rather, we fit it, even though I am driving a dirty, fossil fueled car. The impression of the ancestors and descendants is personal, not impersonally abstract, even if it is imaginative. I can remember my grandparents, can imagine your grandparents from their photos and the stories you told of them, and can hold a rough, imaginative space for those who will come long after me. In all these cases, the impression is to some degree conjectural, my relation both figurative and personal. Mostly, I experience a vivid and meaningful relationship, much as this letter I am writing to you now conveys. For although you are freed from a body turned to ash as I write, I carry your memory so closely inside me that I can write to you now, imagining the year that you were born. I could write to my descendants, too – even to the little one I am about to meet.

Through spiral time, I become more of a related person and this also means, as you well understood, a more *autonomous* one. My sense of myself extends through my relations to become a spiral and not a point. By reaching down and up to ancestors and to descendants who are in my place and I in theirs, I acquire more of a sense of where I come from and of where I am going. I have many "someone[s] to be accountable to"[9] stretching through time, orienting the sense I make out of life. So it becomes harder for me to be egotistical. Spiral time helps prevent me from thinking the world is about me. Moreover, since my kin spiral intergenerationally through a land,[10] my relations have a home that allows them to continue in either direction unless the land that supports them is neglected. So I consider this land that carries us. Spiral time stabilizes my moral judgment and makes me more autonomous, living a life that makes sense to me, rather than one that is arbitrary and wanton.

Spiral time helps me live better inside myself. It eases off anxiety over myself. It helps the tenacious grip of my consciousness become porous and diffuse at the edges, allowing me to phase into a sense of the world in which I both *come from* past others and *fade into* future others. Others speak through me, shoring up my sense of life as transcendent. As when I am in a sudden and rising wind, resting in it, but opened to the outside by it beyond myself and caught in the world's dynamic motion, so spiral time in a quiet and sly way takes me beyond my anxious grip on life.[11] In this way, spiral time is not just moral; it is ethical too.[12]

Spiral time is great. Yet you and I do not live in a society where spiral time is as quotidian as an annual calendar hanging on a refrigerator.[13] We have to internalize something like it in our own way. Asking how I might do this *honestly* took me to "overtone time." I wanted to learn from a tradition, dominated by colonization, that seems to have a way to hold imperialist, capitalist, and industrial society accountable for its planetary wantonness, and I wanted to "learn up" from it in my own way, autonomously.[14]

There is a historical precedent within colonialism of privileging Western knowledge over other forms of knowing and of reducing indigenous knowledge to something that is esoteric or quaint. Such a mentality supports colonialism. It can be found in our nation's history of racist denial of the equal intelligence of different people's ways of being truthful in the world,[15] and it can be seen in attitudes toward the meaning of time!

For instance, according to what I learned in school, there is supposed to be only one obvious way to make sense of time's meaning. Time passes by in a constant flow backward as our lives are arrows flying onward. But this reduction of all the ways of marking time in the world to *one* linear understanding of time erases ways of having relationships.[16]

To lose your world's sense of time is disorienting and violent. Yet the history of knowledge in our country involves other sinister practices, too. Some reduce indigenous knowledge and relationships to medical and scientific categories that interrupt indigenous senses of self and thus *all* relationships.[17] There can be evil in purportedly "objective," administrative classifications.[18] The mindset of these interruptive practices is central to colonialism.[19]

With "learning up," by contrast, one doesn't assimilate indigenous knowledge to pre-existing colonial metrics (e.g., chronological time) and scientific categories (e.g. that spiral time is a mere *fantasy*). One approaches indigenous knowledge as a discourse of equal intelligence that has something to offer to one's own mind and outlook. People are not stupid, although modern education with its endless hierarchies and certifications has implied this consistently.[20] As you well taught me, it behooves someone with a mind to find what others have understood with theirs. Learning up involves opening up a space in one's system of thought to allow a new way of knowing to become imaginable.[21]

Opening one's mind is sometimes referred to as a "decolonial" act.[22] It is the *beginning* of a good relationship in the context of ongoing colonization. Is it the beginning, too, of anthroponomy here where I live? Autonomy in *our* world seems to depend on it. You and I have benefited from colonization living in this so-called "America." We have benefited from colonialism's *mindset*, too, since it hid colonial oppression from us and made the United States of America seem to be a legitimate nation state without a criminal history.[23] It behooves us, children of colonialism, to open up our minds. In fact, we have a moral duty to do so.

Think about how our society raised me. I grew up in school in Ithaca and New Hartford, New York, in a cultural system that ignored decolonization. Decolonization was existentially important for the people of many continents on Earth during the twentieth century as the colonial empires of Europe relinquished their colonies only through violent and non-violent struggles waged by the colonized.[24] Decolonization is still important in places like the United States of America and Canada where indigenous nations live under oppression. The treaties to which they agreed are consistently violated and ignored by Canada and the United States of America.[25] The indigenous nations suffer land-grabs for their lands' resources where corporations interact with local, state, and federal governments in corrupt ways,[26] and the same indigenous nations have had to simultaneously adapt to a dominant society while being violated, defrauded, and betrayed.[27]

In this very same reality, my school system taught me about "the modern world." It implicitly and explicitly contrasted the modern world with "pre-modern" worlds outside or predating it, minimizing their cultures' importance as well as the intellectual development of their people.[28] This minimization included the very same indigenous people and cultures being dominated and exploited. While "modern life" launched onward to a global liberation of everyone in line with the narrative of European history's globalization,[29] not everyone was anyone. Indigenous nations *right next door to where I lived* suffered ongoing domination. Meanwhile, I was taught to be ignorant, including to *ignore the signs* of the moral need for decolonization – for instance, a Senecan billboard that demanded that New York State honor its written word.

This shroud of unknowing into which I was schooled is sometimes called "coloniality" – the mindset that enables ongoing colonization.[30] Coloniality is evil, since it aids and abets colonial domination. Decolonization should have mattered to me growing up, since decolonization challenges evil. New Hartford High School is on land that was once part of the Haudenosaunee Confederation whose treaties have not been consistently respected by the United States of America and by New York State.[31] My high school existed and still exists on morally fraught ground insofar as it is part of a nation state, the United States of America, and its "New York State" which aren't in morally right relations with the Haudenosaunee. All that has been part of the ongoing evil of colonialism.

In no way can such "coloniality" be part of our autonomy as a society. In high school, I should have been taught to honor treatises with the Haudenosaunee as well as to honor the independence of these older nations of my land that are not and do not seek to become nation states. Historians think of "decolonization" as referring to a political process of self-determination struggles by the colonized in the late middle of the twentieth century to obtain independent nation states.[32] But this view assumes the nation state unquestionably! The nation state is a colonial political unit. The Haudenosaunee confederacy has nations in it, but they are not nation *states* and operate on their own terms.[33] So even mainstream, historical views of decolonization in the twentieth century have carried on the mentality of the colonial world system, this time in the unquestioned category of the nation state. I should have been taught enough to question even these historians and to honor the independence of my nation state's older nations.

All of what I should have been taught is subsumed in the unwieldy word "decoloniality." Yet that word is precise and logical. Decoloniality is the unworking of coloniality – for instance, of *that* coloniality that clouded my mind and made me a party to a society predicated on ongoing injustice. Decoloniality precedes and succeeds decolonization, challenging the mindset that keeps colonization going. It seeks to resolve the ongoing dynamics of colonial domination, including its traces, so that formerly or persistently colonized people find themselves to be true moral equals with the self-determination to live their lives in their own ways and to contribute – only if they want or have moral reason to – to a shared culture or society as they see fit, forming relationships that are mutually advantageous, accountable, and reciprocal.[34]

Of course, and unsurprisingly, this sense of decolonial work departs significantly from the dictionary context given to "decolonization," even the context implied by the United Nations where you one day will work. The United Nations is still part of the colonial world order, being fixated on nation states, and our dictionary relies on convention among colonizing, Anglophone cultures! "Decolonization" initially appeared with frequency only by the 1950s as an administrative term of colonialist elites for the process of releasing former colonies from empire, transitioning them to national sovereignty.[35] These uses focused on nation state formation in the international order as the purpose of decolonization.[36] They thus *hid* ongoing colonialism in societies where the nation state was not indigenous.

The dictionary definition of "decolonization" does not really fit Cleveland or the United States of America. Confronting colonialism's effects here is ongoing and ambivalently unrelated to forming a new nation state.[37] At best, nations older than nation states demand resurgence, equality, and accountability,[38] and the inheritors of colonial slavery and exploitation demand truly equal opportunity, restoration and capability, rather than being relied on for cheap labor and discarded as a used-up resource.[39] Confronting colonialism here calls for the process of respecting autonomy and actual moral equality with people who are dominated persistently by settler colonialism or the ongoing effects of colonial histories.[40] It also calls for confronting the mindset of a global thought that suppresses the plurality of social processes constituting the world, a thought that is more present than ever with global warming and talk of the "Anthropocene." I am hoping that this will change.[41]

In any case, we have a moral obligation to make change. "Decolonial" work presses toward good relationships, grounded in autonomy, and that is why I think such an unwieldy word would have ultimately mattered to you if you had lived long enough to read this letter. To be "decolonial," we the colonizers have to be self-critical of everything that has allowed us to ignore colonialism with our "knowledge" that is a power subjecting the colonized.[42] For instance, I must disturb and dissolve the mindset that I was taught promoting modern life as opposed to less developed and less intelligent pre-modern and "savage" forms of life. This takes a lot of reflective work. Yet I should do so gladly and fiercely, since moral respect demands it and since, as you taught me, justice clears the way for everyone's capacity to love.

The point is, decolonial work is part of autonomy in contexts where there is ongoing colonial oppression, and so decolonial work is intrinsically related to anthroponomy as a particular process within it.[43] In settler societies such as ours advancing slowly with each land grab or treatise violated,[44] decolonial work involves the spirit and practice of resistance to the mindset developed alongside colonialism that has persevered from the 1500s to this day in mutating forms.[45] Decolonial work in lands such as ours also involves supporting the political struggle for independence from, and the possibility of truly honored agreements with, the United States of America, the imperial power thus humbled and brought back to moral decency. The point is to open up a plurality of worlds, conceptually and politically, so as to return human relations to their freedom.

Decolonial work depends on what one needs to become autonomous. One's role in confronting colonialism depends on one's location, whether one suffers colonialism's ongoing effects directly, feels one has to suppress one's culture and mind, or enjoys the privileges colonialism's history provides for those outside the reservation and the ghetto. The role is positional. A survivor of colonial effects should prioritize autonomy "from within"[46] in order to resist colonialism's oppression, while an ally who is privileged by colonialism's effects should support the autonomy of the survivors institutionally and intellectually "from without."[47] At the same time, an ally should learn up from the colonized and disrupt his or her own systems of inherited thought.

Given my position, I focused in part on what it is to engage in decolonial work on myself. I wanted to get at the persistent inertia of coloniality in my land, society, and way of life – even my values, mind, feelings, and body.[48] Such self-work is a precondition of supporting full decolonization in lands such as ours where treatises are dishonored and domination occurs in subtle ways behind law and policy and within patterns of social violence.[49]

I found that there are two main ways in which to engage in decolonial work on myself: one related to developing an adequate concept of autonomy and the other related to developing an adequate understanding of relationships. But of course, autonomy and good relationships go together.

First, I learned that *what* I hold to be true, good, desirable, or emotion-producing should undergo decolonial work. For instance, I learned in school and from national holidays and culture that I, as an American, should enjoy my liberty. This meant that I am free to do whatever I want, just so long as I do not interfere with someone else's liberty, and that my life is mine to live, just as everyone's is. Being born is the chance of a lifetime, a chance to exercise my liberty with whatever I "got." The idea that I should be accountable to my ancestors and my descendants is thus illiberal, for that accountability constrains me from doing what I want. I am supposed to feel excited and free when I can do whatever I want and somber and constrained when I make myself accountable to my lineage!

But such a sense of liberty makes it hard to tolerate calls for all of us to reorganize our lives and our shared society enough so that the ongoing presence of oppressive conditions in neighborhoods here in Cleveland and in Lorain where Grandpa used to work as a machinist can create autonomy. I was led to think that everyone in the ghetto and everyone who used to work at the steel mill were free to do whatever they want. Too bad that they languish in poverty or stoke the fire of perceptions of injustice! Too bad that their bodies are broken by hard labor or subjected to carcinogens and other toxins over a long life![50] They are free to do whatever they want. They choose to work and live as they do. They are free to get out and try harder to have a better life elsewhere.

This *what* that I was taught had to go through decolonial work. Such a sense of liberty as I absorbed rationalizes arbitrariness in relation to people past and future. It runs roughshod over the land that joins us. Such a sense of liberty helps us ignore histories of injustice that do not simply end when one generation dies and a new generation is at liberty to live differently. Such a sense of liberty undercuts good, autonomous relationships.

To do decolonial work on this *what*, I had to refine what liberty means. This is a significant reason why I sought *autonomy* instead, sought a life that makes sense to people on their own terms.[51] People of equal intelligence might have a different *word* for that life. But then we could disagree about it, and that would be truer to freedom.

Second, I learned that *how* I hold things to be true, good, desirable, and emotion-producing should go through decolonial work.[52] This is the thing that interested me most when I thought of spiral time. For instance, I learned in school and from national holidays and culture that I, as an American, am an *individual*.[53] I should think primarily of myself as an individual, yes with my family, but not mainly with my community, my lineage and the land that has carried me through time. The way I think should be self-centered. I might think about society, but my question should be, "Where does it make room for me?" I might think about history, but my question would be, "Where does it give me a fresh start?" I might think about ecology, but my question should be, "How can I get what I want out of it?" It's not that there is never a time for such questions, but *beginning* with them is problematic.

How can I relate well to my community, be formed constructively by traditions, be informed by intergenerational lineages of care, and maintain accountability for safeguarding the land in which I live and which has carried me, when I first and foremost subject such things to my self-centeredness, negatively anxious that I won't get to see the light of day? How can I see myself as bound up in social injustice?[54] Thinking of myself individually significantly obstructs coming to terms with histories of colonial oppression, and it significantly obstructs much of that anthroponomy would seem to demand. Thus, the *way* I was when I was taught to be myself called for decolonial work.

My coming to overtone time was part of such work – an opening to a freer existence in relationship, relating through our family system to the older nations of this land in which we live. You were born in Pleasant City, not far from Belle Valley, itself not far from the state of West Virginia. Belle Valley is where your grandparents settled and where your father and mother began life together. It is also where your sisters were born. It was a mining community – a place where the utilities of the Earth were extracted to produce short-term power and wealth for people participating in fossil-fuel intensive, industrial, U.S. capitalism. It occurred on what was once indigenous land in a nation of unresolved, colonialist appropriation.

Your family – my family – emigrated to the nation and then to this land in Southern Ohio from the Spisské region of Slovakia, poor people searching to avoid forced conscription by the Hungarians and the Russians and seeking a life where they might become free of some of the hardships of being poor

Slovaks. Our poor ancestors took advantage of colonialism by benefiting from the opportunities opened up for us within it. We thought it was at liberty for us to do. That was *our* coloniality.

So given our family history and given the moral pressure to work on myself "decolonially," spiral time shook a thought loose in me when I read into it. It was like a flowering branch torn from a tree by a gust of wind. I took this branch home, so to speak, and placed it a crystal pitcher like you used to serve summer-time tea. There in water, the branch was, and as I lived in my home, I would feel it in the air.

One day as I was imagining it, I had a thought of a time that made sense in our tradition, a tradition of song. I heard music in the house humming through the refrigerator's sound within the sound of the wind in the trees outside and around our yard. This multilayered sound reminded me of singing, when that surprising music happens in choir and an overtone appears from voices layered over each other. I called this thought, "overtone time."

We sang in our family. Grandpa Bendik always did before sunrise when he would sing Slovak hymns before going to work in the mines or at the machine-shop. You sang too from the time you were a little girl until you sang out under the stars one night in Modesto, California for thousands of people. In our tradition, song resonates with me. Thus it allowed me to make something of it figuratively so that I can more readily remain related intergenerationally in this world. I found overtone time. Overtone time is the time in which generational lines layer over each other and, working in relation, throw off a new line after the sound, as if a future generation were here with us. In overtone time, I imagine past, present and future generations in a relation that suggests a unique power joining them.

I think of this power metaphorically. It is actually a moral relation, a relation of care. What joins the layered time orientations is moral responsibility to each generation in their shared setting, this land. I keep them in mind in this place and process where I live, inheriting a land that bore them and giving over a land that can bear others up ahead.[55] Overtone time is community time, layered up ahead and far back within the place where I live with the land as the basis of our intergenerational community. It is a time seeking good relationships in autonomous harmony.

That's the story.
I miss you, *and* I am with you in overtone time.
Your son,
Jeremy

End file

Mid-autumn, Shaker Heights, Ohio

It is unseasonably warm, but so is the planet now. I've been trying to understand how I should relate to colonialism. I've been reading primarily about decolonization and the related idea of decolonial work. It's hard in a good way. I don't like living in a lie. Besides, it's my moral duty to do this work.

Reading over the past months, I've been most compelled by writers who hold that the main work of decolonization and of decoloniality is to cultivate reciprocity and moral equality. These authors advocate for establishing moral accountability in relationships autonomously chosen in the midst of colonized and decolonized lands with still persisting colonial mentalities and effects.[56] I've come to see that a precondition of such relationships is opening up disagreement between worlds so that "the plurality of worlds"[57] appears. Then the terms of relationships – what and how relationships even are authentically – can themselves be worked through, chosen mutually.

What should I think of the idea that to respond to colonialism plainly and fairly, I should focus on good relationships in my community, my land, and on Earth?[58] The idea is so simple that it troubles me. Yet the idea presents me with a process. It does not pretend to fix things all at once, but it begins to restore what matters morally as people mutually see fit, responsive to disagreement and to histories of legitimate distrust, hurt, and trauma.[59]

As far as I can tell, good relationships bring in the gamut of the more obvious issues in colonialism and its ongoing effects:

- For instance, I cannot dominate another and have a good relationship. If I do, someone else cannot look me in the eye.[60] Domination is a non-relationship, but good relationships must un-work and account for domination.[61]
- Liberty, too, as I've considered, interrupts good relationships because of its refusal to be accountable with disagreement.[62] But good relationships resolve liberty into autonomy.
- Taking land as liberal property fares even worse, because in thinking that land is mine to do with as I please as long as I do not harm or cost anyone else, I lose the way land is the site of relationships, even being a family of them. Land becomes merely an object with no community accountability. But focusing on good relationships uncovers land as a site and source of relationships above and beyond its fluctuating value in an economy or merely its instrumental value.
- Or should individualism strain relationships insofar as it is puts my focus on me rather than on my relation to myself and my relations with others, good relationships make individualism a matter of being differentiated, rather than being a lone individual.[63] In the context of capitalism where individualism becomes self-interestedness and situations become practical opportunities for my investment of time, resources, or wealth, good relationships, by focusing on community accountability, challenge any economy that isn't for the common good.[64]
- Good relationships also situate me intergenerationally, bringing me beyond short term thinking about my own life.[65]
- Finally, in any good relationship, I cannot see those with whom I develop the relationship as hierarchically less intelligent than me and their ways of

thinking and of understanding the world as already invalid. If I were to do that, I wouldn't be capable of taking their viewpoint seriously. I would disrespect their minds. Good relationships, though, seem premised off of a stance of respect for another's intelligence.[66]

In all these ways, good relationships appear practicable and straightforward ways to relate to colonialism's persistence. I've already seen how important relationships are to autonomy and how autonomy begins authentically only in disagreement.[67] I did not realize, though, how much this idea is a decolonial one until I studied decolonial writers.[68]

When I open up to my reality here in Cleveland, I hear, see, and feel how this community is filled with multi-generational rage and trauma over the ongoing dynamics of colonialism as it has translated into racialized inequality and dispossession. These dynamics are bound up with coloniality but exceed them, since they are more than mindsets. They are corporeal and structure the city over time. There is pain to be processed and social justice to be worked through.[69] Still, a substantial part of the situation of Cleveland in its unworked-through colonialism is unworked-through disagreement about the world that is taken to make sense.[70] This is emblemized by the complacency of flag-owners in my community who enjoy the liberty to make fun out of colonialism in their valuable property, showing their fun off.[71] My community isn't structured by actual moral equality and reciprocity. But working through disagreement and aiming for actual moral equality and autonomous reciprocity would attend significantly to the historical wounds in the history and ongoing life of Cleveland. It wouldn't solve them overnight, but it would point in the right direction – toward autonomy and moral accountability.

My hunch is that I should relate to colonialism's ongoing effects by focusing on good relationships.[72] The people and institutions around me may be structured by coloniality.[73] But from my perspective, good relationships begin with me. I am responsible for them in me, not others.[74] That makes them relatable, plain, life-sized – what I have been seeking. Also, because they appear scalable to the relationships between me and institutions and between institutions and each other, good relationships appear suitable for the planetary situation in which I find myself with its twisted history of injustice and torqued inhuman effects, where it is hard to relate my singular life to the institutional problems driving our planetary situation toward environmental injustice.[75] Good relationships relate me with my ancestors and those who come after me.[76] They extend me over time in a way that stretches toward planetary problems by having me consider the Earth we share over time.[77]

Three weeks later, late autumn, Shaker Heights, Ohio

I turned to colonialism as part of a critique of the Anthropocene, and that led me to see how I can clarify anthroponomy. How can anthroponomy clarify moral accountability as our planetary situation becomes more critical and as the predominant way of thinking about our situation lumps all human beings together as an alienated cause of a problem that actually issues from the colonialist, capitalist, and industrialist world? This question led me to decolonial work. And decolonial work leads me to good relationships.

Over the last weeks, I have not stopped thinking about good relationships, beginning with what they involve and imply. What I've found is that the form by which people appear within them is remarkable. Let me explain.

In colonialism, colonizers approach people either as resources or as obstacles.[78] They relate to the colonized practically, and their rationale is practical – aimed at securing economic wealth, resources (including labor), and geopolitical power. But moral accountability, by contrast, is primarily relational, not practical. The way that people appear in moral accountability is *as people* and not as objects. This mode of appearance is at the heart of good relationships.

It isn't enough to be practical to be moral. Practical people can be ruthlessly immoral. Moral people, first and foremost, relate to people as worthy of respect and consideration.[79] Relating, one cannot see a person primarily or merely as a tool, a resource, or as an obstacle. One must see them as a person.[80] Moreover, the same is true of other forms of life – to relate to them morally is to take their lives personally, not just practically. It may even be to personify them. I think of this as the distinction between things and lives, just as between things and persons.[81]

The shift in logic is profound between these two modes of appearance. It demands that one relate, not simply calculate. It opens up deliberation to something that comes before the ends one should pursue. The name for this *a priori* – this "coming before" – is *being-with* others.[82] When we relate, we mustn't try to do something until we are first *with* each other in the relationship. To prioritize the practical is to torque the relationship instrumentally. Modes of relating are wider and more affective than calculation, formed around anything that is involved in being with others *as people*, not simply fitting others into means-end reasoning *as tools*. It is no wonder that a rich moral life involves dreams, fantasies, songs, touch, and much emotion.[83] These are some of the modes of being with others personally.[84] In its modes, relating with others already involves a plurality of worlds.[85]

Thinking about how relating differs from practical calculation, I realized several things.[86] The main one is that switching from the practical to the relational is not something that colonialism, capitalism, or industrialism can handle. For instance, a slave is a tool. The slave owner can calculate how to use slaves and even how to use them up. The same is true of the wage worker for the capitalist,[87] and the land for the industrialist, "rich" in extractive minerals. But once the slave – or the wage worker – is seen as a person (or the land is seen through moral relations, personally[88]), one cannot ignore the suffering and diminishment of that person (or that scarred, excavated, ecologically decimated land) without losing oneself as a person, too![89]

Moreover, suffering and destruction aren't all there is to consider. There is also the matter of disagreement. When people become either obstacles or means to the resources a colonizer or capitalist wants, it does not seem imperative to respect their autonomy. They can be manipulated arbitrarily. But when we relate to each other as people, we have to find ways to reach an actual agreement in any interaction in which we will be with each other. This agreement is not surface level, and it isn't merely practical. It involves having to be open to the moral and existential reality of the other, which, for the extractive industrialist especially, should include other forms of life.[90]

In these ways, to maintain oneself as a person makes the solely practical orientation to the colonized, the subject of capitalism, and to the extractive resource of industrialism impossible.[91] The morality of relating is a problem for the colonial reliance on practicality, let alone on administrative rules that shield the administrator from personal accountability.[92] The same goes for management of others and over lands in actually existing capitalism and industrialism.[93]

The fact that in relating we employ a different kind of logic than in practical life significantly heightens my sense that responding to colonialism morally should proceed through good relationships. *A logical shift is a source for decolonial work.*[94] The moral thought of relating is not just a disagreement around what is legitimate. It is a disagreement around *how* to consider what is acceptable and to process it. Relating implies a different kind of community order, one formed around familiarity, not practicality, first and foremost. So it would seem to provoke us to approach politics differently, too. Even community politics must become a different thing with it.

Focusing on good relationships fits well alongside what I realized in my search to understand how to engage in community politics. There in the thick of autonomy was relationship. I did not ask whether autonomy was in the service of being practical or whether it was in the service of something else. But I depended on the reality of good relationships to keep autonomy authentic by holding disagreement and fostering trust over time. I had not yet seen the role of good relationships in specifically decolonial work.

Suppose, then, that *relational*, not practical, reason should be the ground within which people seeking to disagree and to become more autonomous as moral equals can develop a personally accountable relationship. *It follows then that how I should engage in community politics depends on shifting the mode and logic of consideration to the primarily relational.* Then politics is primarily the repeated essays[95] by which we work on being with each other in community, rather than practical warfare against competing interests and interest groups![96]

In relating, I try to figure out where we each are and how to meet, not in a practical sense, but as people. Relating makes it obvious that a lot of what we know about people is not knowing them at all. Having met, we are not ever simply *facts* or *acts*, things we've figured out to size the other up or ways we now know how to handle or avoid the other, how to make them fit into our schemes.[97] We are now *acquaintances*. Over time, we become *familiar*. If we disagree and fall into discord, insofar as we do not revert to manipulating each other, we have to work things through as people, not as obstacles to each other's ends. In this way, relating isn't manipulative (from the Latin word for a "handful").[98] It isn't objectifying either.[99]

So when I relate, I consider others in a certain way. I consider them first from the *heart*, so to speak, not seeking to *hand*le them or to grasp them in my *head*.[100] The considerations of relating help us find each other and find ourselves. They help us find each other and be with ourselves as people (rather than being self-objectifying[101]). The considerations of relating are inherently personal, and as such, they help us confront any theoretical or practical logic that is fundamentally impersonal.[102] They reach beneath or apart from the practical to find the personal. In considering relationally, we find each other on our own terms.

A good way to show how reasoning relationally develops a different approach to community politics and a different community in the making is to consider the kinds of community that relating opens up.[103] An important example for the history of both colonialization and of coloniality concerns the concept of the planet itself. A core feature of coloniality and of capitalism is the production of the globe as a single space for practical resourcefulness.[104] Within this, the Earth is understood as to-be-mapped, to-be-plumbed, or to-be-capitalized for conquest, industrial resource, and capital opportunity. It is not understood as lands – places of relating within flows of life in which the Earth involves us.[105]

Yet one can relate personally to lands. This takes knowing the land as one knows family and living with it, not simply on it.[106] The Yellowknives Denē scholar and activist Glen Coulthard speaks of "grounded authority" in this context.[107] The land of the indigenous communities he discusses is in a field of moral relations with the community, or rather, it is a part of the community, and the community is a part of it. The intertwining is configured as a personal relationship for all in the community.[108] "Grounded authority" is the general name for land as a moral authority that renders merely practical uses of it null and void, morally speaking, and that braids practicality into an underlying familiar relationship.[109] The land with which we relate cannot merely be an instrument – or a resource, a source of wealth or capital. It must foremost be a relation. It may even be a kind of extended family.[110]

Being relational in my land, I disagree here, in Cleveland with the colonialism and capitalism that continues to dominate life here. Disagreeing with my community in Cleveland by focusing on the land in which we live as an unrecognized and poorly seen possible relation does seem to crack open the industrial system for which colonialism paved the way and which capitalism created and continues. That system has parceled up the land along Lake Erie as property, where the land has served as a conduit for extractive resources such as gravel, salt, and sand; and maintained vestiges of a century old industrial economy centered around petroleum (*Standard Oil*'s once home), petro-chemical industry (*Lubrizol*'s still home), and metal production (aluminum, once steel). The land dug up as stuff by "American" capitalism was not family to anybody. It was private property *owned* by families. It was and still is pure, practical stuff.[111]

**Mid-November, notes after waking up from a
dream, somewhere off the coast of Canada and
of many, older nations in flight from Utrecht,
Netherlands to New York City**[112]

Note 1. I had a dream about my mother. In it, she was in the hills of Southern
Ohio. She was sitting down, and her back was to me. I must have been little, on
my hands and legs as much as on my feet. I've always loved the smell of the grass
and the earth, and there was grass here, greenish blue and rolling over the hill
down into a valley beyond which were other rolling hills.

The only thing else I remember about the dream was that I was preoccupied
playing. At one point, I looked up, and my mother looked over her shoulder at
me, turning from the view along the hills and valley in which she had been herself
absorbed. She then said, "Where are they?"

Her eyes were teaching and bright.

[. . .]

Then I woke to the thin line of the sun neither setting nor rising off the side of
the plane flying through the night at hundreds of miles per hour in the sub-zero
cold of this far up atmosphere.

[. . .]

On waking, I felt strangely happy, also sad, the grief around my mother's death
permeating everything. The dream made me feel that the world is open. I do not
know how or why.

Outside, the world feels alien – this carbon-intensive airplane rushing ahead of
meaning perilously fast.[113] Is that how I feel about our planetary situation and the
main social processes driving it?

Is the memory-trace of my mother, re-combined in the dream, how I feel about
the possibility of familiar relationships?

And who are "they" in the dream when she speaks?

The dream and the feelings with which I was left provided no intuitive
response. Yet the dream made me think about overtone time again. When I wrote
a letter to my mother this past season, I drew on intergenerational, familial relationships, and I noted that these relationships occur in a place. I wrote about
"overtone time," with the overtone layering over a place. I have not said enough
about that yet.

When, in the dream, my mother asked where I am, she was trying to locate us
in a place in relation to some undefined others, "they." The imagery of the place
near Belle Valley over the hill above it was rich and powerful. The question to me
about "they" was suffused by that place.

[. . .]

Note 2. What is land's relationship to overtone time, that is, to familiarity across
time? How is a place a site of accountability across generations? When generations cannot directly relate, how can places mediate the relationship, if at all?

All relationships are mediated. To relate implies a third between the two who relate. The third is the relationship, the mediation.[114]

Nothing about relating is immediate. In fact, one reason that the considerations by which we reason about relating are essentially locaters is that relating occurs through mediation. The considerations we give each other about where to meet or find another personally are considerations about mediation. "He does not seem to be here; see if you can meet him there, in that way." "Are you listening? Try to listen first." It might seem that such reasons are practical, that is, that they are know-how. But their point is not to do something through the person, but to relate to the person.

The question I have is whether land can be a mediation. What would it be for land to be a mediation, not simply an instrument, a resource, or a merely useful thing?

[End of notes, as the plane was preparing to land]

Early December, Shaker Heights, Ohio

Now it is cold. In three days, the weather is supposed to warm again. Cycles of heat and cold, extremer each few years, quickly move through time now. They are part of the oncoming wave of risk for future generations and the dying of the extant order of life on Earth. With global warming in mind, I've been thinking about colonialism now. Looking toward the future, I've turned toward the past. Sometimes the way forward is the way backward.

I'm starting to develop an answer as to how I should relate to colonialism. It passes through the meaning of land in the broadest sense.[115] I should be a relational person first. I should relate through my place. By relating through my place, I should relate intergenerationally so that there is accountability across time. I think it will take me a while to relate these realizations in writing![116]

How should I relate to colonialism? My autumn answer to my summer's question is strange, because it is not focusing on what I relate but on how I do. The answer I've reached is eerie, because it suggests that many arrangements in the land where I live aren't truly relational. They are bad relationships, because their logic is confused by the coloniality pervading them. I wonder if the emotional ghosts coming from the history of violence that haunts Cleveland and the United States of America are reflections of the bad relationships constituting each?[117]

How I relate becomes the critical thing. Relating in place, a place connects me with both past and future generations through the land. What would be a land that is not mere property by which one counts one's wealth? What would be a land that is not merely an instrument or a resource? What would land mean if through and by it we relate to those to whom we must be personally and morally accountable?

All that I know now is that the meaning of the land would be a kind of moral love. Relating through the land, the land would speak care by our intent.[118] In an overtone reverberating through the land, our intent would be to respect the autonomy of others far up ahead while being accountable for the care we have inherited for our own autonomy from the past. The question would then be about how to show our moral responsibility involved in this place *along with* this place. The question would then be, how should we understand our responsibility and show it in this place?

Notes

1 See my See my "Presentism the Magnifier," IAEP/ISEE, University of East Anglia, June 12–14, 2013.

2 See my "Living Up to Our Humanity: The Elevated Extinction Rate Event and What It Says About Us," *Ethics, Policy & Environment*, v. 17, n. 3, 2014, pp. 339–354.

3 Cf. my "Social Process Obfuscation and the Anthroponomy Criterion," *Earth System Governance Project* annual meeting, Utrecht University, November 2018.

4 E.g. The Dutch East India Company and the (British) East India Company Cf. Vandana Shiva, "Oneness vs. the 1%: Creating Equality in Times of Inequality, Creating Solidarity in Times of Polarization," Human Development and Capability Association annual meeting, University College London, September 9, 2019.

5 I mean "pathology" in the non-medical sense as a "malfunction" of a system. See "Pathology," *Oxford American Dictionary*, Apple Inc., 2005–2017. Cf. John Dryzek and Jonathan Pickering, *The Politics of the Anthropocene*, New York: Oxford University Press, 2019, especially concerning path dependences.

6 See my *The Ecological Life: Discovering Citizenship and a Sense of Humanity*, Lanham, MD: Rowman & Littlefield, 2006.

7 Kyle Powys Whyte, "Indigenous Science (Fiction) for the Anthropocene: Ancestral Dystopias and Fantasies of Climate Change Crises," *Environment and Planning E: Nature and Space*, v. 1, n. 1–2, 2018, pp. 224–242.

8 See the *locus classicus* of this impression in Karl Marx and Frederick Engels, *The Communist Manifesto*, New York: Penguin Classics, 2002, where "all that is solid melts into air."

9 Faith Spotted Eagle, "Traditional Leadership from Mother Earth: Standing Rock and the *Mni Wiconi* Gathering – an Evening with Faith Spotted Eagle (*Ihonktonwan Dakota*)," Case Western Reserve University, October 18, 2018.

10 Their kin also spirals so that my relation to my kin involves their relation with theirs, and so on. The spirals braid and weave thickly across communities and deep into the reaches of time across more than a century.

11 See my *The Wind ~ An Unruly Living*, Brooklyn, NY: Punctum Books, 2018.

12 See my "The Moral and the Ethical: What Conscience Teaches Us About Morality," in V. Gluchmann, ed., *Morality: Reasoning on Different Approaches*, Amsterdam: Rodopi, 2013, pp. 11–23.

13 Since there is no institutional accountability around spiral time thus deracinated, it cannot hold us accountable for intergenerational justice. It cannot deal with the problem of "intergenerational buck-passing," in which my generation has an incentive to live large now and pass the costs on to future generations (and so on *ad infinitum* with each generation until the final end of humans). See Stephen Gardiner, *A Perfect Moral Storm: The Ethical Tragedy of Climate Change*, New York: Oxford University Press, 2012, chapters 1 and 5, esp. p. 36.

However, *institutions* of spiral time, for instance in the Onondaga Nation, neighboring where I spent my childhood, might do so. See Onondaga Nation, "Government," accessed February 6, 2019. One question that remains, however, is whether even "seven generations" is sufficient to grasp the depth of the temporal problem as Gardiner frames it. It depends on how literal the expression is. On the interpretation I've learned, the expression is a stand in for unselfish sustainability. In that case, it would be well-positioned to address Gardiner's framing of the moral problem.

14 "Learning up" is an expression I heard recently listening to Faith Spotted Eagle. She contrasted "learning up" with "learning down." "Learning down" is a way to speak down to people. It puts people in their place and presumes them to be ignorant or stupid. Learning down is deadly, especially when part of colonialism. Spotted Eagle, 2018.

15 On Joseph Jacotot's notion of "equal intelligence," see Jacques Rancière, *The Ignorant Schoolmaster: Five Lessons in Intellectual Emancipation*, translated by Kristin Ross, Palo Alto: Stanford University Press, 1991. Cf. Walter D. Mignolo and Catherine E. Walsh, *On Decoloniality: Concepts, Analytics, Praxis*, Durham, NC: Duke University Press, 2018. The concept of coloniality is attributed to Anibal Quijano. See also, Gayatri Chakravorty Spivak, "Can the Subaltern Speak?" in Rosalind C. Morris, ed., *Can the Subaltern Speak? Reflections on the History of an Idea*, New York: Columbia University Press, 2010, part I; Edward Said, *Orientalism*, New York: Vintage Books, 1978.

16 Cf. Walter Mignolo, *The Darker Side of Modernity*, Durham, NC: Duke University Press, 2011. The problem is how to open up the meaning of time so that natural scientific technology is considered as one among many technologies of time, each with different uses and meanings, each different virtues and vices, depending on the context. Simply ignoring the plurality of temporalities globally is a colonial move when it is tied to reproducing colonial systems that subsist in part by erasing the modes of intelligibility of colonized peoples and anti-colonial struggles.

17 Consider, in this light, the rise of the social sciences as a form of police-state control in the nineteenth century, a project interwoven with the administration of colonized populations. On these structures' persistence today in "benevolent" state administration and "social work" regarding indigenous populations, see Jaskiran Dhillon, *Prairie Rising: Indigenous Youth, Decolonization, and the Politics of Intervention*, Toronto: University of Toronto Press, 2017. On the nineteenth century's development of social science, consider Jacques Rancière, *The Philosopher and His Poor*, translated by Andrew Parker, Corinne Oster and John Drury, Durham, NC: Duke University Press, 2004; Michel Foucault, *Abnormal: Lectures at the Collège de France, 1974–1975*, translated by Graham Burchell, New York: Picador, 2004.

18 Cf. Kristie Dotson, "Tracking Epistemic Violence, Tracking Practices of Silencing," *Hypatia*, v. 26, n. 2, 2011, pp. 236–257 and with Marita Gilbert, "Curious Disappearances: Affectability Imbalances and Process-Based Invisibility," *Hypatia*, v. 29, n. 4, 2014, pp. 873–888; also Susan Neiman, *Learning from the Germans: Race and the Memory of Evil*, New York: Farrar, Straus, and Giroux, 2019.

19 Dhillon, 2016; "Institutional History," *The Carlisle Indian School Project*, accessed February 3, 2019; e.g. "Kill the Indian; Save the Man."

20 Rancière, 1991.

21 Cf. Emma Velez, "Why the Decolonial Imaginary Matters for Women in Philosophy," *Blog of the APA*, January 16, 2019; on "decolonizing the mind," see the *locus classicus* in Ngugi wa Thiong'o, *Decolonising the Mind: The Politics of Language in African Literature*, New York: Heinemann, 2011; Mignolo and Walsh, 2018.

22 Mignolo and Walsh, 2018 and Sara de Jong, Rosalba Icaza, and Olivia U. Rutazibwa, eds., *Decolonization and Feminisms in Global Teaching and Learning*, New York: Routledge, 2019. In this book, I do not take up the possibility that "decoloniality" is a redundant term from the perspective of scholars of "postcolonialism." I find the modifier "post" confusing, whereas I find isolating coloniality within colonialism and relating it to colonization, as I do explicitly in this book's glossary, is somewhat clearer. This is not a decision without its problems, foremost among them my privileging of

my understanding of words over the historical nuances of their histories of discipli-nary use. But I have decided to go with word connections (e.g., "de" + "colonial") over academic history, because it strikes me as more logical and thus accessible to the uninitiated.

23 Cf. Neiman, 2019.

24 Jan C. Jensen and Jürgen Osterhammel, *Decolonization: A Short History*, translated by Jeremiah Riemer, Princeton: Princeton University Press, 2017.

25 Glen Coulthard, *Red Skin, White Masks: Rejecting the Colonial Politics of Recogni-tion*, Minneapolis: University of Minnesota Press, 2014.

26 Shiri Pasternak, *Grounded Authority: The Algonquins of Barriere Lake Against the State*, Minneapolis: University of Minnesota Press, 2017.

27 Dhillon, 2016.

28 Mignolo, 2011.

29 Mignolo, 2011 and my *Solar Calendar, and Other Ways of Marking Time*, Brooklyn, NY: Punctum Books, 2017, Earth thought 354, p. 252.

30 Mignolo and Walsh, 2018.

31 For instance, see *Onondaga Nation*, "Treaties," accessed February 6, 2019, which included the Oneida territory in which I was educated. See the land claim by the Onon-daga as well on the takings around Onondaga Lake in purported violation of federal law. Cf. *Onondaga Nation*, "The Complaint," accessed February 6, 2019.

32 Jensen and Osterhammel, 2017.

33 For studies of political concepts that are not reducible to the nation state in the context of First Nations struggles, see Coulthard, 2014; Pasternak, 2017.

34 Cf. Kyle Powys Whyte, "Indigenizing Futures, Decolonizing Our Lands: Indigenous Methods for Transformation," Beamer-Schneider Lecture in Ethics, Morals & Civics, Case Western Reserve University, October 2017; cf. Bas van der Vossen and Jason Brennan, *In Defense of Openness: Why Global Freedom Is the Humane Solution to Global Poverty*, New York: Oxford University Press, 2018, postscript.

35 "Decolonization," from "decolonize," *Oxford American Dictionary*, Apple Inc., 2005–2017; "Decolonization," *United Nations*, "Global Issues," 2018. On the use of "decol-onization" by administrative elites, see Jensen and Osterhammel, 2017.

36 This is ironic, but not surprising. The United Nations *is* a colonial international order in so far as it takes the European nation state as the unquestioned given and institutes a world order on the basis of it. It is a good example of the expression "coloniality" that Mignolo (2011) uses. That does not stop the U.N. from being used *de*colonially, as when indigenous rights are raised within it, or when using its tools, e.g., human rights.

Similarly if one were able to ask the most famous developer of the expression "decolonising the mind," whether it is surprising that an English-American diction-ary is behind the track of the term "coloniality" or "decoloniality," might he say, if he used these words, that it is unsurprising? English-American is a language of global commerce, itself a form of "coloniality." See Ngugi wa Thiong'o in *Decolonising the Mind: The Politics of Language in African Literature*, Portsmouth, NH: Heincmann, 1986.

37 Ironically, this can be true also of nations recognized by the United Nations as "decolo-nized." Cf. Albert Memmi, *Decolonization and the Decolonized*, translated by Robert Bononno, Minneapolis: University of Minnesota Press, 2014.

38 Coulthard, 2014.

39 Cf. Ta-Nehisi Coates, "The Case for Reparations," *The Atlantic*, June 2014.

40 Coulthard, 2014; LaToya Ruby Frazier, *The Notion of Family*, New York: Aperture, 2016.

41 For instance, see Jamie Margolin, "I'm Not Only Striking for the Climate," *The New York Times*, September 20, 2019. Margolin and a generation of young activists, includ-ing Extinction Rebellion U.S., now link global warming to colonialism explicitly in public discourse. Links accessed September 24, 2019.

42 Cf. Michel Foucault, *Power/Knowledge: Selected Interviews and Other Writings, 1972–1977,* edited by Colin Gordan, New York: Vintage Press, 1980.
43 See Chapter 2. See wa Thiong'o, 1986.
44 Pasternak, 2017; Coulthard, 2014.
45 Jensen and Osterhammel, 2017 and Mignolo and Walsh, 2018, "Decoloniality Is an Option, Not a Mission."
46 Coulthard, 2014. The "from within" includes tradition, community, culture, and one-self. It may for many indigenous people mean the land, equiprimordially. See Pasternak, 2017.
47 Kyle Powys Whyte, panel on the future of environmental justice, EJ20, University of Sydney, November 2017.
48 Cf. Waziyatawin and Michael Yellow Bird, eds., *For Indigenous Minds Only: A Decolonization Handbook,* Santa Fe, NW: School for Advanced Research Press, 2012.
49 Amilcar Cabral, *Resistance and Decolonization,* translated by Dan Wood, Lanham, MD: Rowman & Littlefield, 2016; on social violence and coloniality, see Kyle Powys Whyte, "Settler Colonialism, Ecology, & Environmental Injustice," *Environment & Society,* v. 9, 2018, pp. 129–144; Dhillon, 2016.
50 Cf. Frazier, 2016.
51 Some decolonial thinkers would baulk at the term "autonomy" due to its Enlightenment connotations. See Mignolo and Walsh, 2018, "Closing remarks." But as I've explained autonomy, in Chapter 2, it is intrinsically pluralistic and relational, grounded in moral equality and accountability. These same thinkers support just these things. The problem is a colonial concept of autonomy, rather than a decolonial one.
52 Cf. Charles Larmore, *The Practices of the Self,* translated by Sharon Bowman, Chicago: University of Chicago Press, 2010.
53 Alain Renault, *The Era of the Individual: A Contribution to the History of Subjectivity,* translated by M. B. DeBevoise and Franklin Philip, Princeton: Princeton University Press, 2014.
54 Frazier, 2016.
55 Cf. Mattias Fritsch, *Taking Turns with Earth: Phenomenology, Deconstruction, and Intergenerational Justice,* Palo Alto: Stanford University Press, 2018, esp. the concept of "asymmetrical reciprocity."
56 In Coulthard, 2014; Whyte, 2017; Memmi, 2014, and even Van der Vossen and Brennan, 2018, the focus is on institutions that provide the stability and basis for reciprocity. The nature of these institutions varies. For Coulthard, it is the honoring of agreements by a settler colonial state and the capacity to bring failed agreements to justice through the rule of law (cf. also Pasternak, 2017). For Memmi, it is the bringing to account of corruption in formerly colonized countries mainly and indirectly in former colonizer nations that support the corruption. For Van der Vossen and Brennan, it is the establishment of truly open markets where it is not possible to exploit people and not likely to enter into zero-sum interactions on the market. The latter two thinkers, however, do not consider capitalism as part of coloniality, which Whyte, for instance, does.
57 Mignolo, 2011; cf. Chapter 2 of this novel; Mignolo and Walsh, 2018, "Introduction."
58 I think of my nation as part of this, but through my community, for the time being. Cf. Coulthard, 2014, chapter 3 especially, on complexities of the settler colonial nation state as forcing a logic on community. Also, *"my* land" does not mean that I *own* the land, but that I *belong* to it in relations of care.
59 Cf. Colleen Murphy, *The Conceptual Foundations of Transitional Justice,* New York: Cambridge University Press, 2017 on trusting relationships as a way to transition from structural injustice toward justice; Frazier, 2016.
60 Philip Pettit, *Republicanism: A Theory of Freedom and Government,* New York: Oxford University Press, 1997.

61 Here, a relationship is not simply any way to relate A to B, but involves a dynamic process that will be specified later in discussing "relational reason." It is true that you relate to me with your fist if you assault me, but that is not what could be called a true relationship between people. For one thing, at the first sign of possible disengagement, I will be gone or I will try to subdue you before I then leave.

62 See Chapter 2.

63 See Chapter 2.

64 Cf. Wendy Brown, *Undoing the Demos: Neoliberalism's Stealth Revolution*, Cambridge, MA: Zone Books, 2015; Mann and Wainwright, 2018; However, see Van der Vossen and Brennan's (2018) view of the market as positive-sum interactions. Were this so in capitalism, it would have to be communally accountable. Compare Thomas Piketty, *Capital in the Twenty-First Century*, translated by Arthur Goldhammer, Cambridge, MA: The Belknap Press of Harvard University Press, 2014.

65 Faith Spotted Eagle, 2018; however, as I've noted, it is unclear in this instance whether the intergenerational accountability is far-reaching enough to counter Gardiner's (2012, chapter 5) "pure intergenerational problem," which reaches into distant time. I think it does *only if* we understand the spiral of time occurring in a *land* (or in *land processes of "collective continuance,"* a term from Kyle Powys Whyte) which becomes the site of sustainability. I will discuss this in Chapter 4 of this novel.

66 This would not mean that others can't be *mistaken*. It only means that I cannot dismiss others out of hand. That would be to avoid relationship. Rather, disagreeing, we ought to engage together in considering the world, which may set us both straight.

67 See Chapter 2 of this novel.

68 Mignolo, 2011; wa Thiong'o, 1986; Mignolo and Walsh, 2018, "Introduction," where they speak instead of "relationality."

69 Cf. Claudia Rankine, *Citizen: An American Lyric*, Minneapolis: Graywolf Press, 2014 and "An Evening with Claudia Rankine," Cuyahoga County Public Library, Parma Heights Branch, 23 January, 2019; Conroy Chino and Beverly Morris, *Looking Toward Home: An Urban Indian Experience*, Albuquerque, NM: Create Space, 2011 and "Looking Toward Home: An Urban Indian Experience," screening and panel discussion, Native Cleveland, Case Western Reserve University, April 4, 2018. See also M. Carmen Lane, "*Ken'nahsa:ke / Khson:ne:* On My Tongue, On My Back (Family Tree)," Spaces Gallery, Cleveland, Ohio, July 14, 2018–September 30, 2018.

70 Cf. Neiman, 2019.

71 See Chapter 2.

72 If you will, this "hunch" is my word for political judgment.

73 Cf. Pasternak, 2017 on the state's claims to authority and the legal system's organization around the state and its conceptions of property, land and colonial taking.

74 Compare Levinas's famous statement taken from Dostoevsky's *Brothers Karamazov*, that we are all "guilty," but "I more than others." Levinas takes this to be a *conceptual* point about accountability toward others. See Alain Toumayan, "'I More Than the Others' – Dostoevsky and Levinas," *Yale French Studies: Encounters with Levinas*, n. 104, 2004, pp. 55–66. I don't, however, think the language of guilt helps here. Consider the difference between political responsibility for structural injustice and individual liability for it; Iris Marion Young, *Responsibility for Justice*, with a forward by Martha C. Nussbaum, New York: Oxford University Press, 2010. I am accountable for establishing good relationships on my own basis, but that does not mean that I am guilty of the bad, destroyed, or lacking ones that history gave us.

75 Cf. Gardiner, 2012.

76 Cf. Fritsch, 2018, pp. 19–21, where a familial sense of generational overlap is contrasted with a more extensive notion of overlapping generations. So far, the little I've said of good relationships or show or interfamilial generational relationship seems to fall within the weaker, "familial" definition.

77   See my "Presentism the Magnifier," 2013.

78   Memmi, 2014, p. 20, which makes the point that "neocolonialism" is usually a misapplied word. In the process, Memmi assumes a hard moral characterization of colonialism as exploitation of resources and "theft at all levels."

79   Stephen Darwall, *The Second Person Standpoint: Morality, Respect, and Accountability*, Cambridge, MA: Harvard University Press, 2006.

80   Darwall, 2006 and my *The Wind*, 2018, which discusses Kate Manne's arguments against the bare morality of seeing people as people. Manne conflates a practical sense of a person as an obstacle with a relational sense of a person as someone who deserves consideration. Cf. Kate Manne, *Down Girl: The Logic of Misogyny*, New York: Oxford University Press, 2017.

81   See my *The Ecological Life: Discovering Citizenship and a Sense of Humanity*, Lanham, MD: Rowman & Littlefield, 2006, lecture 6 especially.

82   Mignolo and Walsh, 2018, "Introduction"; Luce Irigaray, *Two Be Two*, translated by Monique Rhodes and Marco Cocito-Monoc, New York: Routledge, 2001.

83   Consider "The Grand Inquisitor" stretch of Fyodor Dostoevsky's *The Brothers Karamazov*, translated by Richard Pevear and Larissa Volokhonsky, New York: Farrar, Straus, and Giroux, 2002. It is a moral fantasy.

84   In addition to Darwall, 2006, see my "Do You Have a Conscience?" *International Journal of Ethical Leadership*, v. 1, 2012, pp. 52–80; "The Moral and the Ethical," 2013; and *Solar Calendar, and Other Ways of Marking Time*, 2017, study 4 especially, "I want to meet you as a person". See also my *The Wind*, 2018.

85   When Mignolo (2011; with Walsh, 2018) speaks of "love" or "care" as the decolonial resistance to coloniality's ideological power, one way to understand him is to say that he is urging that we relate first and foremost, grounded in disagreement that itself expresses the trust and care of authentic relationship, as discussed in Chapter 2.

In *The Darker Side of Modernity*, Mignolo (2011, p. 1) begins by talking about his rereading of Stephan Toulmin's *Cosmopolis: The Hidden Agenda of Modernity*, Chicago: University of Chicago Press, 1992. He points to Toulmin's thesis that the hidden agenda of modernity is the "humanistic river" running behind "instrumental reason" (p. 1) and *contrasts* his understanding of coloniality with such a picture. An obvious question to ask is whether turning to relating to decolonialize practical patterns of coloniality is simply to repeat Toulmin's thesis and thus to re-inscribe a decolonial gesture within a form of coloniality. But note that, like Manne, Toulmin seems unaware of the relational. His book crescendos towards a chapter on *practical* philosophy. The problem, in my terms, would thus be that "humanness" gets re-inscribed within practical reason, whereas it belongs if anywhere in relating, first and foremost.

86   One thing I learned is that the history of philosophy is impersonal to its core, since it has only recently made way for relational reason. All of its morality prior to this recent, post-eighteenth-century emergence of relational reason had to be reduced to practical reason, that is, to the logic of calculation in the attempt to secure one's objectives. Is it any wonder that thinkers such as Kant created eerie moral categories like an "end in itself" given such a tradition or that contemporary thinkers of practical reason cannot quite countenance moral vulnerability within their focus on seeking the good as a practical matter?

87   Cf. Sam Adler-Bell, "Surviving Amazon," *Logic*, August 3, 2019.

88   See my 2006, lecture 5, "Relations Between Humans and Lands."

89   "Relations Between Humans and Lands," lectures 4 and 6, "Rooted in Our Humanity" and "Being True to Oneself."

90   See my "Friendship, Freedom, and Love: Mutuality with Other Species," Human Development and Capability Association Annual Meeting, University College, London, 2019.

91   Consider the transformation of Ms. Ault in Frederick Douglass's *Narrative of the Life of Frederick Douglass*, Dover Thrift Editions, 1995, chapter 3. She cannot remain truly personable and be a slaveholder. To slave, the person must die and be replaced by the dominator. Cf. Pettit, 1997.

92 Cf. Memmi, 2014.

93 See Robert Jackall, *Moral Mazes: The World of Corporate Managers*, 2nd ed., New York: Oxford University Press, 2010, where introducing the counterfactual, "what if the managers related interpersonally?" to the situations studied would unravel the corporate ethos almost immediately, chaotically, and possibly self-destructively for the corporations involved.

94 See Waziyatawin and Yellow Bird, 2012; Spotted Eagle, 2018. However, I am uneasy that we would be in the territory indicated by Ngugi wa Thiong'o (1986), unless we take on Walter Mignolo's distinction between the "content of the conversation" and the "terms of the conversation," e.g., the "rules of knowing." See his contribution in Mignolo and Walsh, 2018, p. 212.

95 Attempts, tries.

96 See the Hobbesian notion of politics in Jason Brennan, *Against Democracy*, Princeton: Princeton University Press, 2016, where politics is a struggle for power that divides us as obstacles to each other, a logic that is practical all the way down and thereby exclusive of relating. But also see a much more sympathetic Russell Hardin, "Democratic Epistemology and Accountability," *Social Philosophy and Policy*, v. 17, n. 1, 2000, pp. 110–112. It too focuses on interest groups. Coloniality is pervasive. See also my *The Wind*, 2018, "Figures of imagination," especially the letter to "militants."

97 Here, I contrast relational reason with theoretical and practical reason. See my "Do you have a Conscience," 2012.

98 "Manipulate," *Oxford American Dictionary*, Apple Inc., 2005–2017.

99 Cf. Emmanuel Levinas, *Totality and Infinity: An Essay on Exteriority*, translated by Alphonso Lingis, Pittsburgh: Duquesne University Press, 1969, esp. its critique of "noesis," that is, of phenomenological intentionality as found in the Husserlian tradition. See also, Jean-Luc Marion, *Reduction and Givenness: Investigations of Husserl, Heidegger, and Phenomenology*, translated by Thomas A. Carlson, Evanston, IL: Northwestern University Press, 1998 on the problems with the phenomenological "object."

How is it that someone can fail to know you even when they know every single thing about you (or so it seems)? It's because knowing a person is not the same thing as knowing a lot about a person, just as it isn't the same thing as knowing how to handle a person. See my "*Ad hominem* address," Seattle University, April 2009, especially the chart at the end of the paper.

100 The "heart" is a sentimental expression, but it gives us a useful contrast with other modes of reasoning, e.g., the practical (the hand) and the theoretical (the head). Perhaps it helps to understand that the heart, on my account, *supports and works through* disagreement, not suppresses it in a fake show of caring or uniformity. See my *Solar Calendar, and Other Ways of Marking Time*, 2017.

It is worth emphasizing that relational reason does not operate *exclusively*, since we are beings who must think through all modalities to live well. The question is one of priority. Which form of reasoning *leads* the way, that is *organizes* the interaction? A *primarily* relational interaction is different from a primarily practical one. See "Do You Have a Conscience?" 2012.

101 See Brown, 2015 on neo-liberal self-objectification.

102 As for instance, colonialism, capitalism, and even industrial mass production-based communism are or were (on communism as ruthlessly practical, see Moishe Postone, *Time, Labor, and Social Domination: A Reinterpretation of Marx's Critical Theory*, New York: Cambridge University Press, 1993; Herbert Marcuse, *One Dimensional Man: Studies in the Ideology of Advanced Industrial Society*, 2nd ed., Boston: Beacon Press, 1991). Communism *was also* part of the "dark side of modernity," i.e., coloniality (Mignolo, 2011).

103 For instance, the rejection of the priority of the practical undercuts the standard social contract rationale for joining in a community, namely, to protect one's practical

interests as unrelated and merely aggregated individuals. Cf. C.B. Macpherson, *The Political Theory of Possessive Individualism: Hobbes to Locke*, New York: Oxford University Press, 2011.

104 Mignolo and Walsh, 2018, "Decoloniality Is an Option, Not a Mission"; see also Neil Brenner, *Implosions/Explosions: Towards a Study of Planetary Urbanization*, Berlin: Jovis, 2014.

105 Fritsch, 2018, "Interment"; Mignolo and Walsh, 2018, "Introduction."

106 Coulthard, 2014, chapter 2, "For the Land: The Denē Nation's Struggle for Self-determination"; see also my 2006, lecture 2, "Moral attention and justice," and lecture 5, "Relationships between humans and lands."

107 Coulthard, 2014, chapter 2; Pasternak, 2017.

108 Cf. Andrée Boisselle's work as part of the project *Making Room for Indigenous Law in Canada: Towards a Reconception of Western Legal Theory*, Trudeau Foundation/University of Victoria, 2008–present, accessed February 3, 2019.

Are we considering another way to think of a land community as Aldo Leopold (*A Sand County Almanac, and Sketches from Here or There*, New York: Oxford University Press, 1968) invoked in canonical environmental ethics? But Leopold's rationale *erased* colonialism. It was part of coloniality. In his book, Leopold spoke of a second Enlightenment that would *extend* dignity to the world beyond humans, eventually to lands as systems ("The Upshot"). This drove indigenous relations to land further into oblivion, as if they never were until an "evolution" of "modern" ethics came along through Europe.

109 Cf. also Pasternak, 2017.

110 Cf. "Relationships between humans and lands," "The idea of an ecological orientation," and "Moral attention and justice" in my *The Ecological Life*, 2006. The last discusses the "Sacred Mother" of the Amungme in Irian Jaya, Papa New Guinea. The mother is a mountain – whose head, the mountain top, was "removed" by the Freeport Mining Company of New Orleans, LA, U.S.A.

111 Cf. "Standard Oil," "Lubrizol," and "The Flats," *Wikipedia*, accessed October 23, 2018; "Aluminum Company of America," *Encyclopedia of Cleveland History*, accessed October 23, 2018, and Courtney Verrill, "12 Rare Photos Inside a Beautiful Mine That's Hidden 2,000 Feet Below Lake Erie," *Business Insider*, May 11, 2016, accessed October 23, 2018. Compare Coulthard's (2014) discussion of the land as abstract property rather than as a field of communal relations.

112 Irigaray (2001) refers to her practice of rejecting airplanes because of their phallic-like violence in the sky. Whatever one thinks of this ethical position, the airplane is a primary mode of globality and as such could be considered, in the lifestyles it supports and with which it is bound up, as a form of coloniality. See, for instance, Jason Reitman, dir., *Up in the Air*, Los Angeles: Paramount Pictures, 2009 and Manuel Castells, *The Rise of the Network Society*, 2nd ed., Cambridge, MA: Blackwell, 2000.

Ironically, the flight here was returning from the *Earth System Governance Project* annual meeting, Utrecht University, November 2018, which was thus linked structurally to industrialism.

113 A year after writing these lines, I began doubting my access to airplane flights, even with carbon offsets. Even carbon offsetting has potentially colonial problems, and of course supporting airplanes when they have no current transition to a new form of sustainable energy is still supporting their forcing of the planet's carbon sinks. See Olúfe'mi O. Táíwó, "The Green New Deal and the Danger of Climate Colonialism," *Slate*, March 1, 2019, accessed September 25, 2019.

114 Søren Kierkegaard, *Works of Love*, translated by Howard and Edna Hong, New York: Harper & Row, 1962; Stephen A. Mitchell, *Relational Concepts in Psychoanalysis: An Integration*, Cambridge, MA: Harvard University Press, 1988.

115 In other words, land, coast, sky above, etc. A place where we and non-humans live together on, below, and above earth, "on" Earth.
116 This essay also spirals, not just time.
117 M Carmen Lane, "Ken'nahsa:ke/Khson:ne: On My Tongue, On My Back (Family Tree)," mixed media construction with black body bag, Kahyonhá:ke / Cleveland, Ohio, 2018. See also the end of Chapter 2 of this novel.
118 Kyle Powys Whyte and Chris Cuomo, "Ethics of Caring in Environmental Ethics: Indigenous and Feminist Philosophies," in Stephen M. Gardiner and Allen Thompson, eds., *The Oxford Handbook of Environmental Ethics*, New York: Oxford University Press, 2017, chapter 20.

Figure 4.1 Colonial family, Connecticut, 1972

4 How should I respond to the "Anthropocene"?

the "land":
stream of life and death
myriad forms of life with their own considerability
– In "land," we live and we die.
– No, we live up to living or we don't.
– Including you and me, "land" isn't about us.
– Love makes scarce sense without it.
~ Written on a napkin, Utrecht, the Netherlands, late autumn

Writing desk, winter, Shaker Heights, Ohio – once land of many nations

The moral thought that I am involving in "anthroponomy" depends on emotional and relational maturity – on the ability to disagree, hold disagreement, and to grow through conflict in time.[1] It is rooted in respect for personal autonomy – for people finding what makes sense to them and not having senselessness imposed on them by others. As part of that respect, the moral thought of anthroponomy depends on the awareness of the "third" in any relationship, namely the relationship itself. Being aware of the relationship one has with others is a part of being aware of one's autonomy. The fact of this third, of the relationship, introduces differentiation into the worlds of those involved in it. If they are to share a world of common sense, they must be accountable to each other and work through their differences. Even if they do not wish to share a world, they must be accountable to each other when they have differences so that another's world isn't imposed on them.[2] Moreover, no respectful person's world can foreclose the possibility of disagreement arising with others or even emerging inside oneself.[3]

This moral thought is opposed to coloniality. Consider my imperial nation state, the United States of America, that erases indigenous claims and culture while being in disarray as to its treatises with indigenous nations.[4] My nation state continues to dominate the indigenous who have been colonized, despite their opposition and what makes sense to them in their communities and traditions. The dominant world of my nation state isn't open to the plurality of worlds existing in its land and arising from disagreement. Its land isn't shaped collectively out of autonomous disagreement within all the communities involved in the land, many seeking independence and reciprocity, living through different ways of being and understanding, including different languages and names for this land.[5]

The erasures are not merely symbolic. Where I live, they come with guns. They emerge with the day as laws, policies, and other social structures maintaining the seeming rationality of a world insulated from its colonization.[6] Consider the ongoing, bitter senselessness to the indigenous nations of this land of my nation state's erasing of the ongoing, historical crime of colonization. The moral thought involved in anthroponomy proposes instead a common sense in this land that is accountable to decoloniality, a world open to worlds.[7]

I still do not know what that looks like yet, nor how to understand my responsibility in bringing it about where I live. But I am looking to anthroponomy as I confront the planetary situation in which most of us live with many different degrees and kinds of vulnerability.[8] In actually realized moral equality where no one is subject to another, anthroponomy would seem to imply a coordination of autonomy across humankind through disagreement. Being a specific organization of autonomy, anthroponomy also would seem to involve a logic – a form of consideration – wherein all social processes become open to making sense to people, non-dominating, and manifest accountability within disagreement.[9]

Moreover, being organized to make systems accountable for the planetary trajectories of social processes across time,[10] anthroponomy would seem to have the goal of sustaining autonomy over time and around the planet. It would seem to find itself in and across the land, water, and air. These fields of life – the more than human world in which we are so fortunate to be a part, whatever the severity of surviving in nature[11] – underlie and overlay everyone and are the scene in which we can live up to our moral responsibilities or fail them.[12]

Thinking about the fields of life in which we can be autonomous, it makes even more sense that I have been focusing on colonialism. Colonialism has a historical relation to expropriating land and to using it as an extractive resource.[13] The history of colonialism has confronted me here in Cleveland in ways that underlay Cleveland's industrialism and capitalism. In Cleveland, I have seen colonialism's orientation toward land there in the gravel mounds and along a river that's often been on fire from the pollutants in it.[14] This same economy of Cleveland has been a part of a global economy that has roared into being all over the planet.[15] Given how much colonialism has been interwoven with the formation of capitalism and its relation to people and to lands as sources of extractive wealth and given how colonialism continues to dominate people in direct and indirect ways, it makes sense for anthroponomy to engage in decolonial work.[16] It makes sense for *my* work to focus on the land in Cleveland.

But how? Understanding my specific responsibility is the issue that concerns me now. This past summer, I came to believe that any community whose processes are shaped anthroponomically should be, paradoxically, grounded in disagreement around the mystery of the world as it emerges.[17] That would seem to hold for societies as well, at least insofar as their failure to be consistent with moral communities would make the societies oppressive, denying autonomy. *The* world is not something we presume. In the midst of disagreement, it's a finger pointing to the cracked open space in our own world's limit, to our world revealing itself as less than total and as unfixed. "The world" is then a sign for the limitation of a given world.[18] The society that can continue in being accountable to its world's limits doesn't avoid the world's mystery as it surfaces in disagreement.[19] How is the world's mystery a part of the land?

I found one way in different senses of time. This past autumn, I began to consider indigenous senses of time as a way to open up the plurality of worlds in Cleveland. I came to realize that the form of consideration I seek to involve in my response to the "Anthropocene" requires an understanding of people as being accountable to each other for the sense or senselessness they confer across generations.[20] I thought of this awareness as reflecting care and accountability which I take to be central features of moral relationships.[21] What I began to wonder is whether the moral and civic responsibility by which we locate thought about justice, among other virtues,[22] can be realized by giving to future generations of people and the more than human world a thoughtful society. This would be one that isn't wanton but is caring, grounded in the autonomy of community developed through good relationships by which it transforms itself in response to the demands of accountability.[23] I took this moralized conception of time to be a form of decolonial work. The question is how it implies that I should relate differently now. How, for instance, should I relate along with the land differently? Opening up "overtone time" confronted me with a mystery.

Yet by the end of autumn, the thought of the land learned up[24] over the past year began to register with me. Odd as this seemed to me at first, the land can be the "third" in a relationship between contemporaries as well as between generations. In the land where I live, I can find the relational reasons to orient my accountability to past and future generations as well as to the more than human world in our planetary situation. Land can be a field of relationships, not just "on" it but also along with it. Even more than being a condition of care,[25] land can constitute aspects of intergenerational and communal care around the mystery of the world in its changing. Its fields of life are morally considerable,[26] and in them we can learn not only to relate to each other thoughtfully, refusing to be wanton with life, but we can learn to relate with the myriad beings who exist with us on Earth now. In so doing, we can live up to our moral nature.[27] The land emerged as a mysterious, but significant relationship saturated with moral relations.[28]

It seems to me that through our relationship with the land much of what is involved in anthroponomy begins to appear.[29] I want to become clearer on the way the land is involved in anthroponomy so that I can orient my life accordingly. To orient my life was my charge some months ago when I began these reflections in a state of negative anxiety and frustration at the risk of becoming paralyzed in my response to our planetary situation. My community exists in a land – a stream of life far greater than any of us human beings wherein we as people relate. Responding to the mystery of the world and the seeming infinity of the cosmos, the society that I've inherited has made – and continues to make – the land into an inhabited world. The society that I've inherited has made – and continues to make – its inhabitance into a place.[30] The place, however, may not live up to the mystery of the world and the moral considerability of life. A society may become closed off and wanton.[31]

Yet the land is a primary mediation between people and the more than human world across time.[32] Given our planetary situation, this land will be changing profoundly within the lifetimes of our children and any further descendants.[33] What would it mean for care to appear in lands, accommodating places and occasioning times for people and the more than human world? What would it mean for such care to become thoughtfulness, living up to the moral considerability of the myriad forms of life in the land and living up to the mystery of the world? How should I respond to wantonness with lands that, in sharp contrast, undoes moral relations and forecloses the time of humankind as well as the extant order of life? This last question is especially pressing as colonialism has made the land an extractive resource, while the capitalism and industrialism that emerged with and on the basis of colonialism are hurling the planetary order of life into mass extinction.[34]

The thing that seems to matter for anthroponomy is that we come out of and return to lands, lands sometimes changed within our lifetimes, appearing and disappearing within environmental conditions on our becoming and passing away.[35] These conditions are not simply material. They include the moral and spiritual meaning of land and the ways in which we have involved land in the meaning of our being, a being that is now changing with the planet.[36] Land expresses our relating or the lack thereof, and is itself a relational field – or the lack of one – with the more than human world. That more than human world in which we are involved, in turn, has its own myriad relationships themselves in flux, especially in the coming future.[37] All of these things – our lives across time, the lives of other beings, even whether we confront our own finitude[38] and are accountable to the world as a ground of disagreement – are morally considerable in some way so that the land is a field of moral relations.[39] The organization of autonomy to coordinate across humankind as anthroponomy should accordingly consider the land we share with myriad forms of life across time and around the planet.[40]

A week later, Shaker Heights, Ohio, once land of many nations

Reading today with last week's reflections in mind, I came upon an article about an art performance in India a few years ago. The article discussed how a group of artists staged a mock trial on behalf of rivers against the latest iteration of a river-connecting engineering project aimed to modernize conditions across India, ostensibly by engineering a drought-resistant, hydrological system. Part of the question raised by the trial was the question of the legal language that could be used to make rivers claimants of legal right. Doing so would depart from a system of law that traces back to British colonial rule and which gives only human agents the right to have rights. Taking in the wide context of this trial, the author wrote:

[W]hat does it mean to form a new language?

To begin, of course, is to decolonize. To talk about accelerated climate change and the possible end of [our agency in] the Anthropocene is to see . . . the colonial histories that first initiated the mechanization of land and water systems, oftentimes fundamentally altering natural processes and topographies.[41] Topography cannot be so easily divorced from how it carries the colonial legacy. As such, the Indian subcontinent reveals a vast and indomitable plane, where water, not only land, saw immense mechanization projects as deployed by a colonial rule that extended itself across what are now several different nation-states. Consider the city of Mumbai, a precarious land mass formed by the unification of seven small islands by the Portuguese occupation of the eighteenth century. Some of Mumbai's most densely populated areas live on the shifting soil of reclaimed land, mere inches above sea level. Terrifyingly, this reclamation still continues, as the city does not have enough available land to meet the demands of its rapidly urbanizing population.[42]

The author connects colonization, land, and our planetary situation. She suggests that the same kind of social processes "mechanizing" the land in pursuit of colonial aims – e.g., extraction of resources or capture of labor – undergird our terraforming of the Earth on a planetary scale.[43] This suggestion is a strong one. One of the things that makes the "Anthropocene" problematic is its nominal – and self-contradictory[44] – relation to human agency. Our planetary situation is *un* intentionally formed.[45] It is not agential.[46] Either this is a disanalogy with colonialism, or colonialism has within it an unintentional logic strongly similar to that of our planetary situation and the social processes driving it.[47] The intentional terraforming of engineering projects mentioned in the article and our unintentional planetary situation driven as a byproduct of social processes aren't analogous.

Still, the author's suggestion appears to be that there is something qualitatively different about the colonial "mechanization" than would be found in pre-colonial engineering. Perhaps this quality holds the key to the relation the author implies between colonization and the "Anthropocene."[48] Humans societies have consciously shaped the land on which they live at considerable scales since at least the dawn of the agricultural age, including across far-reaching empires.[49] The author implies that something makes European colonial "mechanization" qualitatively different. What?

It may be that the logic of European coloniality provides part of an answer.[50] One thing the article conspicuously omits is a specific focus on the relationship of capitalism to land, even while the article is contextualized within a thesis about capitalism driving our planetary situation.[51] As the author no doubt knows, from its beginning in India, "Western" colonialism was bound up with capitalism, a chiasmus that continues through the history of capitalism to this day.[52] The question of coloniality is thus a question involving capitalism's specific ideology.[53] What, after all, is driving the rapid urbanization of Mumbai – the accumulation of people around the accumulation of profit by the wealthy?[54] The globalization of capitalism is, occurring through spatial conditions and social processes that actually produced "India" in the process of colonization and decolonization.[55] British capitalism drove global imperialism in the first place.[56] What more should be said about capitalism's specific logic within colonialism especially in relation to the land?

Capitalism erases the variegated cultural meaning of land in a global consistency of value – the right of exchange as monetization.[57] Through capitalism, land becomes a space of investment flows to return profit. It stops being, strictly speaking, a place of moral relationships.[58] Temporally, too, it becomes an opportunity, not a tradition.[59] In these ways, the logic of capitalism pushes the planet toward becoming a space of flows rather than of morally considerable places.[60] It conceptualizes the planet as a matrix of opportunities rather than a diversity of traditions.[61] Interestingly, these spatial and temporal shifts are driven by capitalism's logic.[62]

Capitalism is interwoven with modern social processes just as colonialism is. Although it is common to point to capitalism as the driver of a political economy that produces the planet as a space of flows,[63] the drivers of our planetary situation include industrialization as found within state communist projects[64] as well as, in the last near half-century, the Information Age's transmutation of all globalized societies, fully capitalist or not.[65] The goal of state communism was not profit, but it was management of an economy of scale.[66] The goal of the information economy involves capital in large part, but it also includes "global connectivity" and liberty as such, especially among information technology idealists and philanthropists.[67] What unifies these processes in driving our planetary situation?

One way to try to answer this question is to ask what each process seeks. We might begin with the goal of capitalism. This is often given in shorthand as "profit."[68] It becomes an end-in-itself.[69] What is important and disturbing about this goal is not the straightforward claim, which Aristotle originally made about seeking wealth,[70] that such a teleology is irrational. Of course, it is irrational that a means has become an end. All capital is, when monetized, is a means. But the important thing is to read the purpose of profit through what it expresses at the level of its *form*. Formally, capital is a pure potential to obtain goods in any place where a market works. It is a right of exchange.[71] On this line of thinking, the point of capital is to *abstract* its owner's agency from any given context so as to acquire a global potential to live as one chooses *anywhere* in a market society, trading freely and generating capital as one can through mutual agreement with others. In other words, the point of capital is a certain, abstract view of global liberty.[72] It is a certain way of life, characterized by negative freedom – license – and scaled as globally as possible.[73]

In this light, state communist projects appear coherently scaled, if not aimed at promoting liberty, whereas Information Age activism appears entirely consistent with capitalism's tacit drive for liberty, even if information activism isn't driven – explicitly – by profit.[74] The thing that joins them all is a formal commitment to scaling production and circulation on a mass – ideally, global – scale. What makes possible, formally speaking, this global scaling of what we do with the land, some ideologies of which advance liberty while others advance productivity?[75]

We should focus, I think, on a different kind of abstraction found within the form of all three social processes: *the relationship to land*. The land is merely an *opportunity* in capitalism; a *resource* for state communist, industrial production; and an *instrumental condition* of information technology's virtual world. In all three social processes, the land is merely a means of some form, a useful part of a human society that abstracts over land, projecting it out within its ends as a practical object in the midst of the social processes actual ends. In other words, one thing that unifies the processes formally is that land is *never in itself morally considerable*. Land isn't relational.[76]

The logic that thus joins modern social processes such as capitalism, industrialism, and even late twentieth-century information activism with coloniality is a specific *land logic*. The logic is fundamentally and originally colonialist. Colonialism was and is predicated off of a fundamentally abstractive, practical logic that understands places as economic opportunities.[77] This understanding then rationalizes the local meaning of the place being disregarded, including the communal worlds around those places, woven of practical and relational meaning. By abstracting land into a merely practical logic, colonialism represses the plurality of relational worlds and so of relational disagreement around moral considerability more generally. It represses such a pluralistic opening up of moral considerability for the sake of practical goals such as domination, control, extraction, and exploitation, but also – seemingly more benignly – sustainable use or even ecosystem services or capabilities.[78] Thus, whether overtly or covertly, colonialism operates through forms of objectification of lands, including their communities thereby.[79]

Given, then, the importance of colonialism's land logic to the main social processes driving our globalized world and our planetary situation, to do decolonial work should begin by critiquing a purely practical approach to land. Decoloniality should thus be grounded in land relations. This is a claim that indigenous scholars have been making for a long time.[80] It is obvious to me that, as a colonizer, I come belatedly to it.

Clarifying the land logic underneath capitalism and linked historically back to colonialism as it emerged alongside early capitalism, I now understand better the focus of the author I read this morning, beginning with her wish for a "new language." Coloniality erases languages that express land relationships.[81] Provoked by a theater piece staging the trusteeship of rivers as having moral claims of their own, the author seeks a "new language" beyond a world wherein rivers show up in court only as resources for people and institutions. Certainly, this search for a new language isn't marked as reversing the erasure of earlier languages, but the question of language still hovers in air.[82]

Need the language be new? Many indigenous societies maintain language of land's considerability as a field of moral relations including the more than human world, taking on the colonial state in acts of decolonizing resistance.[83] There are long and lasting traditions on this Earth that would disagree with a logic of land where the land has no moral claim on us.[84] Yet our planetary situation does involve one new thing: social processes driving the *Earth*'s pathways.[85] This is new in the entire history of life on Earth. Never before have social processes become geological.[86] The planetary scales of this social forcing are new.[87] What "language" of moral responsibility with the land is then adequate to our planetary situation?

To my mind, this question of my responsibility in my planetary situation is new, even as I am located here in Cleveland. The land to which I belong is located in the Earth system.[88] How does it bear that system while being involved within it? What is a relational understanding of that "system"?[89] How might my land's involvement in the Earth's wide reach bear on my moral and civic responsibility across time in our changing situation? What – or where – is my land? Need it be merely Cleveland or colonialist "America"? How *could* it be only or even these, given their land logic? Most importantly, how can the planetary context of where I live differ from being merely a site of value implicated in the logic of global flows, that is, a capitalist and industrialist expression of a colonialist land logic? Questions like these point the way toward my responsibilities.

Two weeks later, Shaker Heights, Ohio, once land of many nations

In Cleveland, where no indigenous nation explicitly asserts a claim to the land,[90] I inhabit a "settlerscape."[91] The settlerscape of Cleveland is so historically repressive that the only "land" to "defend" is a basketball championship – that is, when one is not in the season of the Cleveland "Indians."[92] Cleve*land* shows the wear of the fabrication and circulation of industrial products over a hundred some years. In the last decade, it has become, increasingly, a good opportunity for gentrification's real estate development. This good opportunity that is pleasant Cleveland is, however, troubled. In the words of a Potawatomi scholar, it is still colonial, involving "vicious sedimentation:"

> Vicious sedimentation refers to how constant ascriptions of settler ecologies onto Indigenous ecologies fortify settler ignorance against Indigenous peoples. . . . In historic accounts of fur traders, clergy, and settlers, they certainly attempted to enclose regions such as Anishinaabewaki into settler concepts of nationhood, savage places, and so on. But . . . the colonists nonetheless traveled through these regions and recognized the different Indigenous ecologies operative within those places. . . . Yet, fast-forward more than two centuries.
> [. . .]
> The Midwestern U.S., for example, appears to settlers . . . as endless farming and commercial agriculture, recreational lakeshore, unoccupied parks, vast urban centers, wilderness space, golf courses, quaint towns, military installations, and so on. When settlers even walk onto an Indigenous jurisdiction or nearby a sacred site, there is a good chance that they experience no awareness of any difference from their own lives. . . . [A]ll they can see are settler ways of life. . . . Urban gentrification in Midwestern cities erases any traces of Indigenous origins of the area. Gentrification processes often commodify highly selective memories and legacies of other groups, often people of color, who lived there before the most recent gentrification process.
> [. . .]
> Vicious sedimentation explains why certain allies are unable to advocate effectively for Indigenous peoples. . . . [Scholars][93] have written about

innocence, in which privileged persons feel that their daily actions and aspirations for justice are not implicated in settler colonial domination. Hence, these persons get to feel good about advocating for Indigenous peoples without having to take on the hard work of doing anything that will change the underlying *land-based structures* of domination that secure Indigenous disempowerment. These underlying land-based structures are what made it possible in the first place for the Dakota Access Pipeline – including the process of its construction – to even be something that some people would envision as good.[94]

This multifaceted passage has a lot to say about restoring a relational understanding of land, starting with my specific historical relationships here in the Midwest of the United States of America. The land where I live was ceded to the political economy of the United States of America in the 1795 Treaty of Greenville after the U.S. military dominated a Shawnee led defense at the Battle of Fallen Timbers.[95] The treaty was meant to end warfare between settlers and indigenous nations whose previously existing hunting grounds were invaded by settlers. But this treaty, like all 374 treaties with Native Americans ratified to the U.S. senate to this day,[96] wasn't honored. True to a nation state that violently represses autonomy, its words were not kept.[97]

On this land obtained through the thoroughly non-relational acts of invasion, military domination, and a dishonored treaty of broken words, a system of liberal private property settled in.[98] To this day, that system expresses a colonial logic by which an owner can heedlessly fly a racist flag – a form of hateful speech denying the personhood of others.[99] Moreover, that owner's property is understood as wealth that can be translated into financial quantities. It is a "right of exchange," a pure potentiality of abstract capital.[100] If the owner wants to eradicate the life on his property – provided that it is non-human and does not harm a neighboring property owner or their "rights of exchange" – the owner can. In a space of flow, all qualitative relationships are reduced to the quantitative throughput of a thoroughly commensurable value system, with only (potentially) private-property-owning individuals protected.[101] Unless the owner damages the wealth of other property owners or the rights of other individuals, the owner can exist in their own world. The land here is as non-relational as the owner set upon it.

There is a complex, morally fraught, violent history in this land where I live. As I determined last summer, one thing that I should do is to point to it and to crack open the world wherein I live.[102] This history includes settler invasion, informal and formal warfare, military domination leading to capitulation, forced treaties, dishonored treaties, liberal property rights, arbitrary individualism without moral accountability to community history, public education downplaying the aforementioned,[103] symbolic mockery, racism, and a violently abstract,[104] instrumentalized understanding of land.

The great absence haunting my land is the absence of moral relationships with others and with the land as a morally considerable field in its own right. The great absence is the diminishment of relational reasoning in the life of a society, something that has consequences for how that society approaches its own time, place and interactions. This is my society's qualitative loss.[105]

The land logic of Cleveland is consistent with our planetary situation too. The specific social processes driving our planetary situation have also been hidden by projecting people as alienated from their own agency, lacking in relational strength and trust, and correspondingly being wanton with their very being.[106] That hiding is summed up in the name "Anthropocene."[107] In the absence of morally accountable relationships, modern, industrial production and empty, abstract rights of pure exchange within capitalism drive us onward.[108]

Here, then, is the qualitative problem joining the personal here and now with the geological now and future. The knot between the daily life of my society and the social processes forcing the planet wantonly into paths destructive of our existing biological order is a constitutively practical and reductively non-relational approach to living.[109] My society in daily life lacks morally accountable, trustworthy relationships both locally and with planetary scales in mind. Instead, its social processes are abstract to each fundamentally arbitrary individual and inexorable on a collective scale. Wantonness appears as an epiphenomenon of social processes reflected in an arbitrariness at the heart of people's sense of themselves, relation to community, and grasp of the world.[110] I think that the heart of this matter is the extraction of the relational from the meaning of being a person in our world, beginning with the extraction of land as a relation.[111] That *extraction* appears as *abstraction*.

Sunday afternoon, late winter, Shaker Heights, Ohio, ancestral lands of Native Americans ceded under threat in 1795

The separation of people from lands is at the heart of the *ex*traction of the relational, and it goes to the heart of the matter in coloniality.[112] It binds the fundamental problems of colonialism to the problem of the "Anthropocene." When lands are merely practical objects, they are set over against people.[113] This creates an antagonistic situation between people and lands. Where people live is structured by antagonism, and the antagonism easily bleeds over to other people and other forms of life. I want to call this quality of coloniality *land abstraction*, where the abstraction is from the land. Extracting the relational leads to land abstraction.

Conceptually separating people from their lands secures a world in which only people become potentially relatable, and lands become merely practical. This then allows an entire region of being[114] to exist that is merely practical. Any being that can be excluded from personhood can be shuffled over the line to the merely practical. This includes the indigenous as *savages*, the colonized as *slaves*, poor people as *labor* and *service*, and more than human animals as *factory farming*. In other words, the division sets up a logic – the reduction of a relational field to merely practical logic – that helps to produce "the dispensability . . . of human beings and of life in general."[115]

The lack of moral accountability to any being or field of beings that nonetheless can be prefigured relationally sets the terms for arbitrariness. Shutting down the relational opens up the arbitrary.[116] There is a lack of an authoritative check on the actions of those who are seen as persons or on the behavior and being of those who are related to personally, such as other forms of life and the land as a morally considerable place.[117] The lack of relational reasoning produces a power asymmetry in the moral system of coloniality between people and lands, including all those human beings who are rationalized away as nonpersons, unrelatable. In the absence of accountability, those set up by coloniality as people are at liberty to do what they please and to use as they wish those beings set up in coloniality as *not* being people and so as *not* being relatable.[118]

This is how the system becomes predisposed to wantonness.[119] It is how the system easily overlooks its history of arbitrarily broken words.[120] Understood most generally, wantonness is the arbitrary disregard of the production of senselessness in the worlds of others and the arbitrary disregard of the sense of the world as this emerges in scenes of disagreement. Wantonness follows on a defect in or absence of relational reasoning, that mode of being that is a core feature of being moral.[121] Wantonness is not malevolence. A wanton person does not intend, as their final goal, harm to others or the annihilation of the world. Rather, in the course of seeking their own selfish or self-absorbed aims, wanton people harm others by producing senselessness for them and act as if the world's sense as it emerges in question through disagreement is not considerable. This begins with not seeing others as relatable, as beings who deserve accounting for one's actions in terms that can make sense to them through a process of relating around disagreement.[122] With respect to beings who cannot talk sense we do, including lands when considered on the whole, wantonness involves not even trying to understand how to read them as having a morally considerable life or how to read the land as being a field of morally considerable life.[123]

Wantonness is a vice of selfishness, including of harmful self-absorption.[124] The settlers who invaded ancestral lands prior to the Treaty of Greenville wanted those lands. What they did not consider morally were the worlds of the indigenous and their lives. The settlers did not consider indigenous worlds interwoven morally with land communities. The settlers ignored the sense of indigenous worlds as these emerged in disagreement. Instead, the settlers saw land selfishly as an opportunity for personal liberty or as economic opportunity.[125] They blocked out the world of the indigenous and subjected that world to the colonial nation's industry and liberalism.[126] Eliminating relatability for the sake of a practical logic constitutive of coloniality, the settlers forced the older nations of this land to capitulate. In this, settlers committed – and we colonizers in "America" still reproduce[127] – a generalized, socially rationalized and supported moral crime against both people and morally considerable lands.[128]

There seems to be a connection with our planetary situation in colonialism that increases the precision of the connection found with the "Anthropocene" when considering the colonially imagined, river engineering projects in India and real estate development and land reclamation in contemporary Mumbai. The social processes driving our planetary situation into perilous paths seek to produce wealth or development and to reproduce societies that are clearly unsustainable.[129] What they are not accountable for are the harms and senseless perils they produce on a geological scale. They currently do not institutionally anticipate these harms in an accountable manner, including facing up to the threat of a radically changed world that will drive masses of people to senseless trauma or death.[130] The social processes and social organization forcing our planetary situation display wantonness, lacking institutional capacities and social processes grounded in relational reasoning. Once again, the knot between colonialism and our planetary situation appears as the prefigured wantonness, the constitutive arbitrariness, of colonial modernity, which represses the relational as the core moral dimension of being people with good relationships.[131]

A basic way to articulate care, or even love, through lands is by confronting wantonness with land predicated off of the relational disregard of land and of the historical peoples who have been forcibly driven from it. For there can be no care, let alone moral love, without basic accountability for a history of violence.[132] We are a long way off from good relationships – and with no guarantee of "reconciliation" – even *with* such accountability.[133] But going to the core moral defect inside historical crime and responding to the depth of wantonness with people and their lands is a beginning form of responsibility against the production of traumatizing and deadly senselessness in the past and its reproduction now and in the future on a planetary scale.

The core moral defect of colonialism and of our planetary situation is structured by what I called *land abstraction*. Lands recede as background practical objects or as antagonistic objects to be handled, and people appear unrelatable, turned into objects to be manipulated, used up, or eliminated. Anyone who has inherited coloniality and its arbitrary liberty is thus responsible for confronting wantonness through advancing a different view of being related, including a different view of freedom than propertied liberty, such as the autonomy in community that I realized through my reflections this past summer.[134] This implies understanding myself as relatable first before being practical[135] and as living in lands that are relatable with people who have suffered the traumas of violation and non-relationship. Such an understanding gives me a personal place to start being responsible in the colonial world in which I've been raised.[136]

Take this land in which I live, this land in Shaker Heights, ancestral land for the older nations colonized to this day by my nation state. This land was lived in – better, lived *with* – before me and long before my society. To remain committed to relationship-in-disagreement, that is to the autonomy-in-relationship I've come to realize is important for considering others, I need to seek the senses of the land that have been violently covered over by my arbitrary nation, including especially the ways in which the land has been more than merely, or solely, practical. These senses of the land involve the ways in which my settler nation's history *broke the words of the land* as the land made sense in the worlds before settler domination. These words of the land are the ways in which the land was a relatable field, reflected in the sense made of and found with it. When the settler world arbitrarily and wantonly took these lands, faking treaties, and maintaining a heedless power to abuse the words of the indigenous, it didn't enter into the disagreement of worlds that is part of autonomy-in-relationship. Rather, it imposed its world and relegated the words of the land to superstition as it did violence to indigenous societies and to the lands here, meting them out as industrial resource and other forms of capitalistic, economic opportunity.

For me to be responsible now, here, in this place called "Shaker Heights," I need to crack open the given world of the United States of America, Ohio, *Cuyahoga* (!)

County, the City of Shaker Heights, and consider the lost words of the land as a relational field including the peoples of the land who were and are woven into it.[137] This field has been minimized or erased for practical purposes, including the peoples living within it who have been treated as mere objects to move around, indoctrinate, or murder in the realization of colonial aims. How has the land been morally considerable and whom did my society violate, whose sense did it erase, to repress the different worlds of the land? How were words – and so *worlds* – broken, and what should I be doing to contribute to making my society begin keeping words – and so worlds – from out of agreements reached on the basis of deeper disagreements?

Here, a new meaning in anthroponomy appears out of the land, so to speak. Anthroponomy should involve confronting broken words.[138] The qualities of its relationships should involve the responsibility to confront wantonness at its arbitrary, promise-less heart. We should come to terms with the relational meanings of our world, including the histories of violence that they carry. At the same time, we shouldn't assume that land must belong to one culture or that the meanings of ancestral lands are the only worlds of this land. The land is larger, more planetary and immemorial than that. The land is a stream of life and of death of myriad forms of life and it exceeds us with the same mystery as the cosmos. *Relatability must be restored* in and with this land as a first step to anthroponomy.[139]

A week later, "the traditional homeland of the Lenape (Delaware), Shawnee, Wyandot Miami, Ottawa, Potawatomi, and other Great Lakes tribes (Chippewa, Kickapoo, Wea, Piankashaw, and Kaskaskia),"[140] a once relational field now viewed practically

I should not be surprised that my responsibility within persistent colonialism involves a change in my orientation toward the world.[141] Immersed in coloniality and hurtling into massively wanton, planetary, environmental change, I can see that anthroponomy, like autonomy, must be a qualitative process, not a state. It should be a kind of relation to living, one that seeks the qualitative relationships that I have been realizing over the past two-thirds of a year. The only way I can see a situation such as the "Anthropocene" being remotely acceptable is if it involves anthroponomy – or some such thing of multiple names and worlds.[142] Otherwise, the sense of our planetary situation will be an imposition on what makes sense to people and will not reflect the orientation of non-domination and the sharing of sense through disagreement in relationship.

In this light, I must form an orientation of relational reasoning toward lands as relational fields, beginning with the land as relatable. I should thereby open up the considerability of cultural fields involving the land in more than practical use. Only with such consideration can the historical and ongoing violence of structuring the land solely as use and as property, a property originally taken under violence, become apparent.[143] Only in such consideration can it become possible to see the sense in the colonized worlds that have been repressed and are to this day oppressed.

Reflecting on the mock-trial art project on behalf of rivers, the author I read a month ago suggested that coloniality and the situation named the "Anthropocene" are continuous. The question is always how. Yet from my wish to develop a relatable response to the overwhelming and depressing situation of the world in which I live,[144] I can sense how a decolonial response to the land in which I live should be continuous with a morally satisfying response to the wantonness of the social processes driving Earth's geology into a perilous state shift.[145] I know that the logic of land abstraction joins them. Now I must orient my life toward the planetary dimensions of our situation more explicitly. From the standpoint of seeing the land as relatable, I must understand how I should respond further, expressly, to the "Anthropocene."

The broken word that is the "Anthropocene"[146] is linked by coloniality to a land logic viciously sedimented over layers of broken words. These broken words of my land continue wantonness with people and with lands. The broken words include broken treatises, but also the promises implicit in treatises to respect the sense of other worlds as held by moral equals. This deeper level of broken words is then subject to the deepest broken sense, which is to see others and their worlds as mere practical obstacles to be manipulated, eliminated, or absorbed. There are, then, three levels of broken words in my land, three levels of bad relationships joining me here and now with my planetary situation.

It seems obvious to me that my moral responsibility is to work to restore these three layers of broken words. The layers of violence in them are profound and compound. They move from treatises violated to moral equality refused to, at bottom, the personal reality of others and their human and more than human families[147] treated as so much obstruction to the selfishness of a still imperial system caught up in the wantonness of actually existing, global capitalism on a planetary scale. Against such multi-layered violence, my responsibility is to confront broken words on every level, restoring moral relations.[148] As a morally accountable person, I must confront the violence of broken treatises, the violence of rejecting the moral equality of others, *and* the violence of seeing others as mere things to be used or done away with. With all of these things, I go deeper into the "Anthropocene" as a planetary condition haunted and driven violently onward by coloniality.

Through the work of this past month, I have realized why it is crucial here and now on our planet that, when we hear broken words in colonialism, we demand that they be honored, that treatises be kept, that people be morally considered despite the law that may fail to consider them, and that no view of the world be tolerated that eliminates in its logic the relatability of the cosmos in the mystery of the world.[149] These are my specific responsibilities and how to show them.

Through them, too, I have begun to enter into a mystery. The situation of this entire planet, Earth, is wrapped up in land abstraction. Land abstraction is vicious to its core. It is wanton. Cracking it open, what world lies beneath? The broken word – "Anthropocene" – expresses a lie circumventing the world about a hundred thousand million people who would never have caused its situation.[150] We have *all* been subjected to colonialism's land logic terrorizing our futures in the names of capitalism and industrialism. A world of broken sense shattered by land abstraction fragments our relation to the world. I need to know how I should respond explicitly to the "Anthropocene," and I finally have an approach.

Notes

1 Cf. my "Environmental Maturity," *Social Theory and Practice*, v. 29, n. 3, 2003, pp. 499–514 and Susan Neiman's *Why Grow Up? Subversive Thoughts for an Infantile Age*, New York: FSG, 2015 and *Moral Clarity: A Guide for Grown-Up Idealists*, revised ed., Princeton: Princeton University Press, 2009; see also my "How to Disagree: Experiments in Social Construction," *Blog of the APA*, July 12, 2018a; "How to Do Things Without Words: Silence as the Power of Accountability," *Public Seminar*, June 28, 2018b; "Democracy as Relationship," *e-Flux Conversations*, April 30, 2017; "Beyond Gestures in Socially Engaged Art: Community Processing and *A Color Removed*," *Public Seminar*, September 6, 2018c; and *The Wind ~ An Unruly Living*, Brooklyn, NY: Punctum Books, 2018d.

As these essays have developed, I've situated the discussion of relationships and "maturity" in terms different than the social condition of "arbitrarianism," to which I refer sometimes as "the Wild West resurfac[ing]" (in 2018b). In "This conversation never happened" (*Tikkun*, March 26, 2018e), I comment:

> People grow up by learning to speak to themselves and by learning to speak with each other. In my interpretation of American life today, arbitrarianism interferes with our ability to speak with each other: "I don't have to listen to you." The root cause is a failure of accountability so profound no one trusts in it. . . . It is a failure that affects social life generally in the fear that no one is safe and that no rule is trustworthy. The failure appears to be of accountability *as a norm*. But this is to open up the social field to only arbitrarianism.
>
> [. . .]
>
> Democratic governments begin with *us*. So that is where the answer must arise. We should restore accountable relationships wherever we can, starting with and from ourselves, anywhere we go, anywhere we work, and anywhere we protest. We should begin with our families and our friends, where it can be so easy . . . to avoid the uncomfortable discussion because it feels unsafe.
>
> You see, the deeper problem is that of not being able to speak to oneself, to trust in oneself. Arbitrarianism is insidious. Accountability begins inside. "I don't want to listen to you" can also be applied within me, whenever I do not want to account to myself or be in relationship to the profound loss that has made life so distrustful that I think that only what I want to do here and now matters. The dismissal of each other is linked to the sense that there is no justice, safety, healing or hope to be had for the damage that we do to each other and which has been − or could be − done to us.

2 Cf. Stephen Darwall, *The Second Person Standpoint: Morality, Respect, and Accountability*, Cambridge, MA: Harvard University Press, 2006; also his *Morality, Authority, & Law*, New York: Oxford University Press, 2013.

3 There could be a case in which a community temporarily foreclosed disagreement *as an agreement* so that relationships might be restored. One would think of this as internal to the community and justified by the priority of good relationships for allowing people to work through disagreement (see Chapter 2). To foreclose disagreement with people external to the community seems to me more problematic, provided that disagreement emerges because of a challenge about shared life. To not hear the disagreement when life is troubled in its sharing appears to disrespect the moral autonomy of others and the moral equality that establishes the space of relational autonomy.

4 Cf. Kyle Powys Whyte, "Settler Colonialism, Ecology, & Environmental Injustice," *Environment & Society*, v. 9, 2018a, pp. 129–144.

5 Cf. Catherine E. Walsh in Walter D. Mignolo and Catherine E. Walsh, *On Decoloniality: Concepts, Analytics, Praxis*, Durham, NC: Duke University Press, 2018, p. 17ff.; also, Sereana Naepi, "Pacific Peoples, Higher Education and Feminisms," in Sara de Jong, Rosalba Icaza, and Olivia U. Rutazibwa, eds., *Decolonization and Feminisms in*

Global Teaching and Learning, New York: Routledge, 2019, p. 18ff.; Waziyatawin and Michael Yellow Bird, eds., *For Indigenous Minds Only: A Decolonization Handbook*, Santa Fe, NW: School for Advanced Research Press, 2012, p. 2ff. Regarding the "cultural bomb" as found in Ngugi wa Thiong'o, *Decolonising the Mind: The Politics of Language in African Literature*, New York: Heinemann, 2011.

6 Kyle Powys Whyte, "On Resilient Parasitisms, or Why I'm Skeptical of Indigenous/ Settler Reconciliation," *Journal of Global Ethics*, v. 14., n. 2, 2018b, pp. 277–289; Jaskiran Dhillon, *Prairie Rising: Indigenous Youth, Decolonization, and the Politics of Intervention*, Toronto: University of Toronto Press, 2017; Waziyatawin and Yellow Bird, 2012.

7 Could one argue that disagreement is itself a world? Would it simply be a world that allows the plurality of worlds and in this way is open to the hermeneutic arrival of a world as something to be interpreted and re-interpreted? Cf. Hans-Georg Gadamer, *Truth and Method*, translated by Joel Weinsheimer and Donald G. Marshall, New York: Continuum, 2004. But what would this look like in a set of social structures, policies, communal practices, etc.?

8 Of course, some of us do not live with much vulnerability at all and have benefitted immensely from industrialism, capitalism, and imperialism. But even the global political-economic systems supporting the privileged are fragile to planetary scaled disruptions and to global political economic turmoil such as a pandemic.

9 Accordingly, anthroponomy would seem to challenge the most privileged especially, since they profit from – or are at copious liberty within – the systems of inequality established by imperialism and capitalism especially. Anthroponomy would seem to cause much trouble for actually existing capitalism, imperialism, and colonialism.

10 Will Steffen et al., "Trajectories of the Earth System in the Anthropocene," *Proceedings of the National Academy of Sciences*, v. 115, n. 33, July 2018, pp. 8252–8259.

11 Cf. Benjamin Hale, *The Wild and the Wicked: On Nature and Human Nature*, Cambridge, MA: MIT Press, 2016.

12 Cf. Breena Holland, *Allocating the Earth: A Distributional Framework for Protecting Environmental Capabilities in Law and Policy*, New York: Oxford University Press, 2014; and her "Environment as Meta-Capability: Why a Dignified Human Life Needs a Stable Climate System," in Allen Thompson and Jeremy Bendik-Keymer, eds., *Ethical Adaptation to Climate Change: Human Virtues of the Future*, Cambridge, MA: MIT Press, 2012a, chapter 7. Holland does not consider the environment as a condition on living up to our moral responsibilities, but does conceptualize the ways in which it is a condition on human dignity.

13 Shiri Pasternak, *Grounded Authority: The Algonquins of Barriere Lake Against the State*, Minneapolis: University of Minnesota Press, 2017.

14 "Cuyahoga River Fire," *Ohio History Central*, accessed September 26, 2019.

15 On "planetary urbanization" as the trajectory of capitalist geography, see Neil Brenner, *New Urban Spaces: Urban Theory and the Scale Question*, New York: Oxford University Press, 2019.

16 For my first rough and largely uninformed statement of this, see my "Decolonialism and Democracy: On the Most Painful Challenges to Anthroponomy," *Inhabiting the Anthropocene*, July 27, 2016.

17 See Chapter 2. Cf. Alessandro Ferrara, *The Democratic Horizon: Hyperpluralism and the Renewal of Political Liberalism*, New York: Cambridge University Press, 2014; Mignolo and Walsh, 2018, seeming opposites.

18 In Kantian terms, it is the "X" by which a given conceptual scheme becomes problematic, and in Hegelian terms, it is the "negative." See Immanuel Kant, *Critique of Pure Reason*, translated by Norman Kemp-Smith and Howard Caygill, New York: Palgrave Macmillan, 2007; G.W.F. Hegel, *The Phenomenology of Spirit*, translated by A.V. Miller, New York: Oxford University Press, 1977.

19 Cf. Kyle Powys Whyte's term "collective continuance" in his "Indigenous Philosophies of Environmental Justice: Braiding Traditional Knowledge, Resistance and

Decolonization in the Anthropocene Epoch," International Association of Environmental Philosophy annual meeting, Memphis, TN, October 21, 2017, and Whyte, 2018a. See also his "Collective Continuance," in G. Weiss, A. Murphy, and G. Salamon, eds., *50 Concepts for a Critical Phenomenology*, Evanston, IL: Northwestern University Press, in press.

20 I mean "generation" in the conventional sense, not in the sense of Stephen M. Gardiner, *A Perfect Moral Storm: The Ethical Tragedy of Climate Change*, New York: Oxford University Press, 2011, chapters 1 and 5.

21 Cf. Matthias Fritsch, *Taking Turns with the Earth: Phenomenology, Deconstruction, and Intergenerational Justice*, Palo Alto: Stanford University Press, 2018.

22 That is, in which second-personal considerations shape the logic of possible or actual address. Cf. Darwall, 2013, part I, "Morality."

23 Cf. John S. Dryzek and Jonathan Pickering, *The Politics of the Anthropocene*, New York: Oxford University Press, 2019, esp. the term "ecological reflexivity." See also *Earth System Governance: Science and Implementation Plan of the Earth System Governance Project*, 2018, accessed November 27, 2018, "Contextual Conditions: Transformations."

24 That is, *from* Glenn Coulthard, *Red Skin, White Masks: Rejecting the Colonial Politics of Recognition*, Minneapolis: University of Minnesota Press, 2014, chapter 2, "For the Land: The Dene Nation's Struggle for Self-Determination," which I first read in 2016 and, more recently, from Pasternak, 2017.

25 As it would be, for instance, following Holland, 2014.

26 See my *The Ecological Life: Discovering Citizenship and a Sense of Humanity*, Lanham, MD: Rowman & Littlefield, 2006, especially lectures 2–5 and 10.

27 *The Ecological Life*, 2006, lecture 6 especially, "Being True to Ourselves."

28 Cf. Jean-Luc Marion, *In Excess: Studies of Saturated Phenomena*, translated by Robyn Horner and Vincent Berraud, New York: Fordham University Press, 2004.

29 Cf. Walsh, "Insurgency in Decolonial Prospect, Praxis, Project" in Mignolo and Walsh, 2018, p. 35:

> The struggles for and on territory and land as base and place of identity, knowledge, being, spirituality, cosmo-vision-existence, and life, have long organized the collective insurgent praxis of ancestral peoples, identified as Indigenous, Afrodescendent, or Black, and sometimes as peasant or *campesino*. Such struggles are lived today through Abya Yala [the Gune name for the land called "North America" in colonial geography; *explanation mine*] in both the South and North.

30 1. Cf. Steven Vogel, *Thinking Like a Mall: Environmental Philosophy After the End of Nature*, Cambridge, MA: MIT Press, 2015. The environment is inhabited, which is to say that "it" takes on our meaning as "it" is lived in, becoming the meaning, to us, *of* our living in it. Even more precisely, we could say that the very idea of living *in* an environment is part of our understanding of it. It could also be lived *on*, and, moreover, it can also be made into a meaning *for* us, as when a dominant world imposes a meaning on the environment that is opposed to our living in it. The subtlety here is to understand that *the* world, which is unknown in disagreement, becomes also *a* world that we might understand as "nature," "environment," "Pachamama" (cf. Mignolo, 2011), "the wild," etc. To be most precise (and confusing), then, even *inhabitance* is a conceptual form of "inhabitance" (for that "it" – "X" – that is "inhabited").

Where I depart from Vogel, however, is in his *totalizing* social constructivism. Everything environmental, he thinks, is socially constructed. Yet to my mind, this totalization of the figure of *construction* represses what resists us enough that we *must* construct something. In other words, it represses the moment of negation that would give us a reason to *assert* or *make* anything at all out of "X" that is void yet "which" stirs or confronts us. That is odd, since Vogel in the same work recognizes the way in which every artifact, much less non-human nature, is "wild" – that is, open to contingency and anomaly. See Vogel's rejection of Adorno's "negative dialectics" around this point in *Against Nature: The Concept of Nature in Critical Theory*, Albany, NY: SUNY Press, 1996. On philosophizing relationally from the "negative" – which in that work I call "the void" – see my *The Wind*, 2018d.

Yet Vogel's work on the "practical ideality" (he calls it "social construction") of the environment is very helpful. It makes us aware of our world. Yes, it is incomplete without registering the negative dialectic in which we engage relationally with the more than human world resisting our constructions. But it also moves twentieth-century Anglophone environmentalism beyond the untenable pre-Kantian duality that has plagued it historically and which still plagues popular discussions of environmentalism often among activists and people seeking environmental lifestyles. Vogel's work is a kind of philosophical corrective to the dualistic history of alienated romanticism. On this last point, see for instance my critical review of David Rothenberg's highly creative but still dualistic *Always the Mountains*, "The Seven-Eleven," *H-Nilas*, May 2003.

2. Cf. also Zev Trachtenberg's work on his site *Inhabiting the Anthropocene*, which he approaches through niche construction theory – e.g., "The (civic) republican niche (Part 1)," September 25, 2019. Regarding "habitability" Trachtenberg – in "Why 'habitability'?" June 30, 2014 – writes:

> [T]he Anthropocene idea also points to a crucial recursiveness in the idea of habitation, which is linked also to the ancient idea of "second nature." That idea, attributed to Cicero, expresses the idea that human beings themselves construct the basic conditions of their lives – the physical environment they inhabit (so fundamental and given as to be, in effect, nature) is the result of their transformation and exploitation of primordial materials and forces (so distinct from those as to be, in effect, a second thing). The Anthropocene seems like the fulfillment (for better or worse) of the idea of second nature: human activity has re-worked not just this or that landscape, but the entire planet. We recognize the recursiveness of habitation when we recognize that habitability is its *product*, not its precondition. That is, a place is made suitable for living by the activities of its inhabitants, who deliberately alter it to become suitable for the kind of habitation they require. This phenomenon, which has been called "niche construction," appears to be quite general across organisms. Human beings are exceptional only for the extensiveness of their niche construction activities; that they engage in them is a core feature of their kinship with other forms of life.

Much of this applies to the land as I understand it. Yet in so far as Trachtenberg also uses construction metaphors that figure the land *practically*, my constructive criticism of Vogel's philosophy also applies to Trachtenberg's. These useful constructivist theories need to be modified to include *relational* reason, beginning with "negation."

31 I first used the concept of wantonness in "The Sixth Mass Extinction Is Caused by Us" in Thompson and Bendik-Keymer, 2012a, chapter 13 and in "Species Extinction and the Vice of Thoughtlessness: The Importance of Spiritual Exercises for Learning Virtue," in Philip Cafaro and Ronald Sandler, eds., *Virtue Ethics and the Environment*, New York: Springer, 2010, pp. 61–83.

32 See Fritsch, 2018, chapters 3–5 as soon as he reorients the "immemorial" in Derrida and the "origin" in Mauss toward the Earth as a gift prior to any human exchange.

33 Steffen et al., 2018; Mark Hertsgaard, *Hot: Living Through the Next Fifty Years on Earth*, New York: Houghton Mifflin Harcourt, 2011 – written in part to and for his daughter.

34 See my and Chris Haufe's "Anthropogenic Mass Extinction: The Science, the Ethics, the Civics," in Stephen Gardiner and Allen Thompson, eds., *The Oxford Handbook of Environmental Ethics*, New York: Oxford University Press, 2017, chapter 36.

35 Fritsch, 2018, chapter 5, "Interment." See also Holland, 2014.

36 See my 2006, lecture 2, regarding "ideological" relations with lands, and lecture 4, regarding involving the more than human world in our sense of our own being who we are. "Ideological" there means simply involving an idea of what the land is and

means, not simply relying on it to survive with basic material necessities. I now think of ideology differently, however, as the use of ideas to *manipulate* others (that is, as a vice, a misuse, of ideas). Today, I would speak about "moral and spiritual" relations with lands, which is more concrete.

On the change to our being in our planetary situation, see especially the introduction to Thompson and Bendik-Keymer, 2012a on "adapting humanity" by changing our conception of who we are. See also Dryzek and Pickering, 2019, p. 35, on the notion of "reflexivity," "a capacity to *be* something different when necessary, rather than just *do* something different."

37 Cf. Coulthard, 2014, chapter 2, "For the Land: The Dene Nation's Struggle for Self-Determination." This presents an ironic contrast with the slogan common to Cleveland Cavaliers professional basketball, "Defend the Land!" See also Pasternak, 2017 who acknowledges actual land defenders.

38 See, for instance, the project begun by Jean-Luc Nancy in *The Experience of Freedom*, translated by Bridget MacDonald, Palo Alto: Stanford University Press, 1994.

39 See my 2006, lecture 5 and Whyte 2018a.

40 Cf. Dryzek and Pickering, 2019 on "ecological reflexivity."

41 Arundhati Thomas is referring to the thesis of Jason Moore that the "Capitalocene" has already superseded the "Anthropocene." Moore's argument has the advantage of pointing out the implications of the word *anthrōpos*, used in the dominant terminology under proposal for our planetary situation. The "Anthropocene" implies human agency. His point is that what is driving our planetary situation is *not* human agency and its modes of responsibility. Rather, it is the impersonal logic of capitalism, which, at a systemic level far beyond even individual societies, drives societies and our lives. But I think such a distinction is premature. We should include capitalism within the colonial matrix of power and so focus more generally on coloniality and the colonial matrix of power, of which capitalism is a part. Once we do, we see that the "Anthropocene" has *already* alienated human agency, as I suggested in Chapter 1. It has done so as an expression of coloniality, as I suggested in Chapters 2–3. See Jason W. Moore, *Capitalism and the Web of Life: Ecology and the Accumulation of Capital*, New York: Verso, 2015; Jason W. Moore, ed., *Anthropocene or Capitalocene? Nature, History, and the Crisis of Capitalism*, Oakland, CA: PM Press, 2016.

42 Skye Arundhati Thomas, "A Garland of Rivers: On the Trial-Performance, 'Landscape as Evidence: Artist as Witness,'" *e-Flux Conversations*, June 12, 2017. The author elsewhere notes that the idea for river connecting projects in India first arose as part of British colonial rule. Multiple colonial powers dissected – and intersected in–India as part of the colonial world system.

43 For instance, via "planetary urbanization." Cf. Neil Brenner, *Implosions/Explosions: Towards a Study of Planetary Urbanization*, Berlin: JOVIS, 2014.

44 See Chapter 1 of this novel on the reification of human agency in the "Anthropocene."

45 See Dryzek and Pickering, 2019, esp. pp. 27ff., "Pathological path dependency in Holocene institutions;" Geoff Mann and Joel Wainwright, *Climate Leviathan: A Political Theory for Our Planetary Future*, Brooklyn, NY: Verso, 2018; my "'Goodness itself must change' – Anthroponomy in an Age of Socially-caused, Planetary, Environmental Change," *Ethics & Bioethics in Central Europe*, v. 6, n. 3–4, 2016, pp. 187–202; Moore, 2015; Dale Jamieson, *Reason in a Dark Time: Why the Struggle Against Climate Change Failed – and What It Means for Our Future*, New York: Oxford University Press, 2014; Gardiner, 2011.

46 This is a conceptual mistake in the otherwise brilliant article by Dipesh Chakrabarty, "The Climate of History: Four Theses," *Critical Inquiry*, v. 35, n. 3, 2009, pp. 197–222 – a mistake that has propagated across research and cultural domains. *Pace* Chakrabarty's claim of humankind becoming a geological agent, we should emphasize that specific social processes within specific societies across humankind have become geological *un-agents*. On this line of thought, see Mann and Wainwright, 2018, esp. chapter 5; my 2016; and a much earlier piece of mine, "The Strange Un-Agent of Our

Species, Our Collective Drift," ISEE/IAEP Conference, Nijmegen, Netherlands, 2011. For the conceptual point by which an agent is to be understood under the logic of intentions and ends, see G.E.M. Anscombe, *Intention*, Cambridge, MA: Harvard University Press, 2000.

47 Mann and Wainwright, 2018, speak of our social *conjuncture* to express the way in which a logic outside of direct agential control forces a situation by a necessity that is systemic, even if the system itself is, at bottom, contingent. On systemic forcing, see also my "The Sixth Mass Extinction Is Caused by Us" and Steven Vogel, "Alienation and the Commons," in Thompson and my 2012a, chapters 13–14. Yet on the fundamental contingency of social processes, see Jacques Rancière, *Disagreement: Politics and Philosophy*, translated by Julie Rose, Minneapolis: University of Minnesota Press, 2004.

48 On the importance of the qualities of relationships to different forms of "collective continuance," see Whyte, 2018a.

49 Cf. Vogel, 2015; Trachtenberg, 2014. See also Matthew Ridley, *The Origins of Virtue: Human Instincts and the Evolution of Cooperation*, New York: Penguin, 1998, esp. on *unintentional* terraforming by way of killing megafauna at a distance, thereby affecting regional ecologies and providing the first biogeological signal of our potential to trigger a mass extinction cascade through our technology.

50 Mignolo and Walsh, 2018; Walter D. Mignolo, *The Darker Side of Western Modernity: Global Futures, Decolonial Options*, Durham, NC: Duke University Press, 2011.

51 Namely, Moore's 2015 thesis.

52 Moore, 2015. See also the (British) East India Company, *Wikipedia*, September 23, 2019.

53 I mean "ideology" precisely here as I use the term: ideas in the service of manipulating people.

54 Cf. Mike Davis, *Planet of Slums*, New York: Verso, 2007.

55 Cf. Manu Goswami, *Producing India: From Colonial Economy to National Space*, Chicago: University of Chicago Press, 2004 which emphasizes nationalism as well, consistent with coloniality stretching beyond decolonization. On capitalism, see also Mann and Wainwright, 2018, chapter 5; of course, Brenner's (2014) shift from the logic of *globalization* to the logic of *planetary* urbanization (shifting the focus from the globe to the planet) signals an important conceptual shift from coloniality contained in human time to coloniality spilling over into geological time and hence to the situation ill-named as the "Anthropocene."

56 See Ellen Meiksins Wood, *Liberty & Property: A Social History of Western Political Thought from Renaissance to Enlightenment*, New York: Verso, 2012, chapter 1.

57 See Kōjin Karatani, *The Structure of World History: From Modes of Production to Modes of Exchange*, translated by Michael K. Bourdaghs, Durham, NC: Duke University Press, 2014.

58 Cf. Manuel Castells, *The Rise of the Network Society*, 2nd ed., Cambridge, MA: Blackwell, 2000 on the space of flows verses the meaning of places.

59 Cf. Marx and Engel's claim that capitalism vaporizes traditions in the name of profit, as sketched in their *Communist Manifesto*, translated by Gareth Stedman Jones, New York: Penguin Classics, 2004.

60 Cf. Brenner, 2014.

61 Cf. Mann and Wainwright, 2018, chapter 5.

62 Mann and Wainwright, 2018, chapter 5, p. 101, rehearse Marx's abbreviation of capitalist production and circulation – "M-C-M'" ("Money–Commodity–Monetary profit"). From a moral and ethical point of view, the thing that is distinctive about Marx's abbreviation is the *telos* of monetary profit – i.e., capital gained – rather than communal autonomy or any common good, among possible ends.

63 As Mann and Wainwright, 2018, do.

64 Moishe Postone, *Time, Labor, and Social Domination: A Reinterpretation of Marx's Critical Theory*, New York: Cambridge University Press, 1998. One of the virtues of Postone's classic work in the Marxist tradition is his awareness of the extent to which

state communism is subject to the same field of critique as capitalism due to its form of production. See also Dryzek and Pickerting, 2019, who qualify the role of capitalism in producing the "Anthropocene."

65 Cf. Dryzek and Pickering, 2019, pp. 13–14.

66 Postone, 1998; however, China now *is* capitalist.

67 Castells, 2000.

68 Strictly speaking, capitalism's goal is the abstraction, *capital*.

69 Mann and Wainwright, 2018, chapter 5, are consistent with this answer in their discussion of capitalism. It allows them to understand how the planet as such becomes a management problem following on the contradictions created when a pure abstraction of value driving ever greater exploitation of opportunities pushes the planetary order into state shifts such as found within global warming's most dire risks.

70 Not capital *per se*. Aristotle, *Nicomachean Ethics*, translated by Christopher Rowe, New York: Oxford University Press, 2002, 1096a5–1096a10, esp. "Wealth is clearly not the good we are looking for, since it is useful and for the sake of something else." I.e., wealth cannot be an end-in-itself, or a "final" end.

71 Karatani, 2014.

72 Cf. Bas van der Vossen and Jason Brennan, *In Defense of Openness: Why Global Freedom Is the Humane Solution to Global Poverty*, New York: Oxford University Press, 2018.

73 On license, see Chapter 2.

74 The Information Age is very much enabled and structured by profit-seeking, however, whatever the activists dream.

75 That this issue is deeper than the critique of capitalism – *pace* Mignolo, 2011 – see Postone's (1998) Marx, and plausibly Max Horkheimer and Theodor W. Adorno, *The Dialectic of Enlightenment*, translated by Edmund Jephcott, Palo Alto: Stanford University Press, 2002; Herbert Marcuse, *One Dimensional Man: Studies in the Ideology of Advanced Industrial Society*, 2nd ed., Boston: Beacon Press, 1991; Martin Heidegger, "The Question Concerning Technology," in *The Question Concerning Technology, and Other Essays*, translated by William Lovitt, New York: Harper Torchbooks, 1977.

76 Vogel, 2015, includes the metaphysical resources to understand *how* land can be projected (he uses "constructed," but the specific active conceptualization made by practices here is of a *projection*). Yet as I've noted, he assumes that such projection must be "practical." This is due to some ambivalence regarding relational logic in his work, where he is suspicious of environmental philosophy that "listens" to a supposedly unconceptualized – "immediate" in Hegelian terms – nature outside of anything human. He is right to steer away from reified immediacy that suppresses the practical conditions of its conceptualization. However, as I have been arguing, the *relational* is a discrete logic that, ironically, is *precisely what capitalism*, for instance, *suppresses*. One might then argue, in the spirit of a philosophical friend, that Vogel's work internalizes some degree of alienation itself in its reduction of the environment to practical logic. To be fair, in other parts of his work, Vogel does seek the relational space of deliberative democracy. Bringing out the role of relational reason in a critique of alienation would help with this goal, too.

77 E.g., as extractive ones, for the metals used in information technology and economized in capitalism.

78 On the last, see Holland, 2014. Holland is an opponent of colonialism. But contemporary colonialism can continue untouched when merely practical and seemingly benign categories for its objectification are used.

79 When a community relates to land interpersonally, to reduce the land to ecological capabilities or sustainable value is still to erase the community through its erasing its land relations.

80 See Coulthard, 2014; Pasternak, 2017. My thesis is not novel in the least and is instead, at best, further hermeneutic integration of this important and widespread indigenous

thesis into reflection on "ethical" (or as I would not say more precisely, *moral*) adaptation to climate change. See also Walsh and Mignolo, 2018. Thompson's and my, *Ethical Adaptation to Climate Change*, 2012; and also Mann and Wainwright, 2018, specifically with respect to the "adaptation of the political." My addition to this latter discussion is, most precisely, to drive home that the adaptation of the *moral* is crucial too, specifically, through relational reason and, more specifically, through a relational land logic. In other words, by continuing Thompson's and my project decolonially, I also seek to modify Mann and Wainwright's adaptation project through work on moral relations, which is lacking in their project.

81 Cf. Ngugi wa Thiong'o, *Decolonising the Mind: The Politics of Language in African Literature*, New York: Heinemann, 2011; Coulthard, 2014; Pasternak, 2017; also, personal communication with Andrée Boisselle, April 2016, referring to Sto:lo land relations ordering society as "laws" in forms of speech and expression that are nonetheless erased by Canadian legal forms.

82 Recovering indigeneity is, however, fraught differently in India than it is in the land where I live. Hindu fundamentalists claim indigeneity against Muslims, for instance. Indigeneity thus becomes a weapon of domination. Laura Hengehold reminded me of this, as did Ananya Dasgupta. See also Martha C. Nussbaum, *The Clash Within: Democracy, Religious Violence, and India's Future*, Cambridge, MA: Harvard University Press, 2008.

83 Pasternak, 2017.

84 Whyte, 2018a; Pasternak, 2017.

85 Steffen et al., 2018.

86 Dryzek and Pickering, 2019. In previous work, I held the view that the megafauna extinctions of roughly 12,000 years ago were part of the social processes by which our entire planetary geology has been altered. However, I now think this is incorrect. The megafauna extinctions caused by killing at a distance following on the invention of atlatls and spears (cf. Ridley, 1998) were substantial, but they did not count as initiating a mass extinction cascade or even extinction rates alarmingly higher (up to 1000 times, cf. Martin Gorke, *The Death of Our Planet's Species*, translated by Patricia Nevers, Washington, DC: Island Books, 2003) than the historical range. Thus, the prehistorical, indigenous extinctions of megafauna, e.g., on the land called by colonialists "America" which made the giant sloth or the wooly mammoth go extinct, were not geologically forcing. What we can say is that they provide the first signal of the technological power of humans.

87 Dryzek and Pickering, 2019, Preface.

88 Dryzek and Pickering, 2019, Preface, pp. 15–16 especially; Frank Biermann, *Earth System Governance: World Politics in the Anthropocene*, Cambridge, MA: MIT Press, 2014.

89 The issue here is that natural scientific categories are often expressive of coloniality. See Mignolo 2011. Yet indigenous knowledge need not be hostile to science, and to assume so is ironically colonialist and condescending. See J. Maldonado et al., "The Story of Rising Voices: Facilitating Collaboration between Indigenous and Western Ways of Knowing," in Michèle Companion and Miriam S. Chaiken, eds., *Responses to Disasters and Climate Change: Understanding Vulnerability and Fostering Resilience*, Boca Raton: CRC Press, 2017, pp. 15–26. The question of the names of the Earth is an implied subject of my "Social process obfuscation and the anthroponomy criterion," *Earth System Governance Project* annual meeting, Utrecht University, November 2018; and in my "Autonomous conceptions of our planetary situation," International Society for Environmental Ethics annual meetings, H.J. Andrews Experimental Forest, Oregon, 2019; also delivered at International Association of Environmental Philosophy annual meeting, Duquesne University, Pittsburgh, PA, 2019; and Earth System Governance annual meeting, Oaxaca Mexico, November 2019.

90 As Susan Dominguez taught me, my urban area "resides on the ancestral lands of Indian tribes of the Northwest Territory ceded in the 1795 Treaty of Greenville" (email correspondence, February 26, 2019).

91 Whyte, 2018a, p. 140 includes the notion of "settlerscapes" in what he calls "settler sedimentation," a term that to my mind implies the phenomenological settling of meaning into a given. See Edmund Husserl, *The Phenomenology of Internal Time Consciousness*, translated by James S. Churchill, The Hague: Martinus Nijhoff, 1964.

92 This is a reference to the slogan "Defend the land" as a way to rally Cleveland Cavalier fans in 2017 around their 2016 National Basketball Association championship. The "land" is Cleve*land*.

93 Eve Tuck and K. Wayne Yang, "Decolonization Is Not a Metaphor," *Decolonization: Indigeneity, Education & Society*, v. I, n. 1, 2012, pp. 1–40.

94 Whyte, 2018a, pp. 138–139, emphasis mine.

95 See "Treaty of Greenville," *Ohio History Central*, accessed March 3, 2019. See also Michael M. McConnell, *A Country Between: The Upper Ohio Valley and Its Peoples, 1724–1774*, Lincoln: University of Nebraska Press, 1992.

96 Personal communication with Susan Dominguez, March 12, 2019.

97 See Whyte, 2018a regarding the unrecognized treaties around the Dakota Access Pipeline, treaties unrecognized by the U.S. legal system when adjudicating property claims concerning the pipeline's passage. That *words* are not kept is especially important from the standpoint of autonomy, which depends on making *sense* together, including coming to *agree* on a shared world out of the sense found *together* in it. (Note, too, that this agreement is a far cry from a social contract, which is voluntarist, whereas relational autonomy depends on the grounding of the world opened up through disagreement.)

98 Tuck and Yang, 2012, p. 5, write:

> Settler colonialism is different from other forms of colonialism in that settlers come with the intention of making a new home on the land, a homemaking that insists on settler sovereignty over all things in their new domain. Thus, relying solely on post-colonial literatures or theories of coloniality that ignore settler colonialism will not help to envision the shape that decolonization must take in settler colonial contexts. Within settler colonialism, the most important concern is land/water/air/subterranean earth ("land," for shorthand, in this article.) Land is what is most valuable, contested, required. This is both because the settlers make Indigenous land their new home and source of capital, and also because the disruption of Indigenous relationships to land represents a profound epistemic, ontological, cosmological violence. This violence is not temporally contained in the arrival of the settler but is reasserted each day of occupation. . . . In the process of settler colonialism, land is remade into property and human relationships to land are restricted to the relationship of the owner to his property. Epistemological, ontological, and cosmological relationships to land are interred, indeed made pre-modern and backward. Made savage.

My understanding of coloniality departs from Tuck and Yang's in that I think that a roughly settler colonial land logic *extends* throughout the capitalist and industrial world system. Their description is *generally* apt. In this way, settler colonial land invasion, re-inhabitance, and ideological reformation show us what is fundamental to our planetary situation. It is quintessential, disclosing what is happening to the globe in the so-called "Anthropocene" (which then becomes another name of erasure, like "America").

Tuck and Yang, I expect, would be wary, however, of generalizing their point as it might dissipate responsibility to deal with ongoing colonization here and now in "North America," for instance. But see my argument in Chapter 5 of this novel. Generalizing some such claims as Tuck and Yang makes disrupts the global system of legitimacy and involves decolonization here and now.

99 See Chapter 2. Note Whyte's (2018a, p. 138) claim, "Gentrification processes often *commodify highly selective memories* and legacies of other groups, often people of color, who lived there before the most recent gentrification process" – emphasis mine. Chief Wahoo is not even directly *a* memory but a distorted, racist fantasy that bears, in its repression of settler violence, memory as a *stain*. It is also a symbol of culturally appropriated mid-twentieth-century gentrification.

100 Karatani, 2014.

101 In his article, Whyte (2018a) focuses on the "insidious loops" of settler ideology that rationalize – i.e., in *a settler* world, apparently *justify* – the further erasure of prior land treaties and understandings, then the expropriation of that land for its resources or locational value in a capitalist economy or a state's political agendas. These loops undo the qualities of relationships, according to Whyte. For instance, they (apparently) excuse people from moral accountability to disagreement or from taking seriously the "decolonial option" (Mignolo, 2011) of the plurality of worlds (violently repressed) outside a given world, beginning with the world involving non-humans, for instance, or the world of those who are masked by racism.

102 I share this responsibility with every other colonizer.

103 See, for instance, *Study.com*'s account of the Treaty of Greenville where the "Americans" thought it was their right" t0 invade the ancestral lands and the "Indians" "resented" it (accessed March 3, 2019)! The website caters to U.S. "middle school" "American" history.

104 If the abstraction tears up moral relations, it is necessarily violent, uprooting moral considerability and opening the way to abuse and unjustifiable killing ("unjustifiable," because moral considerability has been eliminated).

105 Referring to the qualities of relationships in a society, see Whyte, 2018a.

106 The social processes are often hidden *by* the defenders of future generations and the vulnerable, even. One way to interpret the slow-moving debate between Gardiner (2011) and Fritsch (2018) is over Gardiner's morally ambivalent *appropriation* of the logic of *homo economicus* – the self-interested individual – as a *moral* (!) premise of his analysis of global warming. Fritsch rightly pushes on the givenness of this assumption, but to my mind does not go far enough to show the ways in which *Gardiner actually repeats coloniality* in his earnest and otherwise greatly important attempt to defeat one of modern society's effects, i.e., global warming. The assumption of *homo economicus* by Gardiner is part of the problem.

107 See Chapter 1.

108 Mignolo, 2011.

109 Whyte, 2018a emphasizes that global warming is not just "for all humanity" in the same way. The "our" of "our ecological orders" is not uniform. Rather, indigenous people – e.g., the Inuit – are among the most vulnerable to its effects now and in the near future. However concerning the destruction of our *biological* order, see my and Haufe's 2017.

110 Cf. Vogel, 2012.

111 To treat the land as an extractive resource demands *first* that its relational meaning be extracted, that is, taken out of the land's meaning. Colonial epistemologies of the "backwards" indigenous "belief system" support such an extraction. See Mignolo, 2011; Tuck and Yang, 2012.

112 In Vogel's (2015) terms, this separation is a specific *construction* of the environment. I agree, with the caveats I've made previously about his theory lacking relational reason. The land has to be *seen* – which for Vogel would also mean *practiced* – as a field of life that is non-instrumental, that is, which is *relational* (morally considerable and not merely calculable) in order to see the problem of separation that I'm marking. So, if it helps, please try to so see or practice relationships with the land. See my 2006, lecture 5 on some quotidian practices.

113 The description of an object being set "over against" the "subject" is a frequent expression in twentieth-century phenomenology. It follows from the etymology of "object" – "thrown in the way of." See "Object," *Oxford American Dictionary*, Apple Inc., 2005–2017.

114 I.e. *an* ontology.

115 Mignolo, 2011, p. 6. See further (pp. 11–13) where Mignolo traces out the colonization of nature through a changed epistemology of it, beginning with Father José de Acosta in 1590 and Sir Francis Bacon's *Novum Organum* in 1620, where " 'nature'

was there 'there' to be dominated by man" (p. 11). Mignolo traces the further trans-formation of nature in early capitalism into the Industrial Revolution, by which time " 'nature' became a repository of objectified, neutralized, and largely inert materiality that existed for the fulfillment of economic goals of the 'masters' of the materials" (pp. 12–13). Note the practical logic and the elimination of relational, moral consid-erability in lieu of a domain of "objects" (not "persona") and "materials" (i.e., means for some goal).

116 One might speak even of ego in a moral void. Cf. my *The Wind*, 2018d.

117 On the role of personae and personal identifications in more than human relations, see my "Do You Have a Conscience?" *International Journal of Ethical Leadership*, v. 1, 2012b, pp. 52–80 and 2006, esp. lectures 2, 4, 5, 8, and 10.

118 Lands *are* relatable, using personification and other forms of relational reason. See my *The Ecological Life*, 2006, lectures 4 and 5. See even more Pasternak, 2017. On the setting up of persons vs. nonpersons in colonialism, see also Mignolo, 2011, chapter 1.

119 It is worth comparing Dryzek and Pickering's (2019) on the lack of anticipatory fore-sight in the face of our planetary situation with Gardiner's (2011) Pure Intergenera-tional Problem, predicated off of the lack of relatability across distant generations. To what extent is the *predisposition* to lack of anticipation in the existing global systems (or "fragmentation of agency," cf. Gardiner) a result of the reduction of the relational to the practical as a feature of coloniality? Yes, Earth System science is needed to anticipate state shifts on a planetary scale as Dryzek and Pickering correctly assume. But this need comes out of nowhere in an environment where the moral reasons for anticipatory foresight are already problematized by viewing future generations as unrelatable next to existing, arbitrary, self-interested maximizers of their own wealth, e.g., *homo economicus*. Again, Gardiner's otherwise great work is still clinging to the problem. See Fritsch, 2018, too on his critique of this economic assumption that I take to be quintessentially colonial.

120 Recall the discussion of doing decolonial work on the "how" of a colonial order in Chapter 3's opening letter.

121 Cf. Mignolo and Walsh, 2018.

122 Cf. Emmanuel Levinas, *Otherwise Than Being or Beyond Essence*, translated by Alphonso Lingis, Pittsburgh: Duquesne University Press, 1998.

123 Pasternak, 2017; my 2006, lectures 2–7.

124 Cf. my 2012a and 2010, where wantonness is understood as *thoughtlessness*. There, I considered people *without* selfish aims as potentially wanton. I no longer think this is morally accurate. Genuinely unselfish people who cause harm without their knowl-edge are not wanton. Being unselfish, they should respond morally to the harm they cause, and they may realize that their having caused harm may express other vices (such as a lack of understanding or critical foresight), but they are not wanton.

125 Cf. Mignolo, 2011, chapter 1 again.

126 No more bold expression of this is found in the motto emerging within the Carlisle Indian *Industrial* School (emphasis mine) that one should "Kill the Indian" to "save the man." "Institutional history," *The Carlisle Indian School Project*, accessed Febru-ary 3, 2019. Although it is merely a coincidence, I find the red "C" of the explicitly culturally genocidal Carlisle Indian School eerily similar in graphic design to the "C" of the Cleveland Indians baseball team. *I wonder whether from the standpoint of an indigenous person living here*, the two graphically reinforce each other? "Another red C?" If so, the graphical environment would display another version of the settlerscape Whyte, 2018a, delineates. How would this interact with intergenerational trauma and its triggers? Would one be seeing Carlisle walking around on that cap? On intergen-erational trauma, consider Faith Spotted Eagle, "Traditional Leadership from Mother Earth: Standing Rock and the *Mni Wiconi* Gathering – an Evening with Faith Spotted Eagle (*Ihonktonwan Dakota*)," Case Western Reserve University, October 18, 2018, and Dhillon, 2017.

127 I.e. "collectively continue," Whyte 2018a, 2017.

128 The crime against land is not seen in the otherwise correct perceptions of the crimes against humanity, sometimes even genocide, committed against Native Americans. It is *highly colonial* that crimes against land are not seen as part of crimes against humanity unless the land is seen as a practical condition the undermining of which harms people systemically. See Peter Singer, *One World: The Ethics of Globalization*, New Haven: Yale University Press, 2002. Singer's work here, despite its chapter "One Atmosphere," is thus still part of the problem, as Gardiner's (2011) was, due to unworked-through coloniality.

129 Mann and Wainwright, 2018, chapter 1; Dryzek and Pickering, 2019, pp. 4–5 esp.

130 Dryzek and Pickering, 2019; Gardiner, 2011, chapter 7 on the "Global Test."

131 The work I have read on coloniality (see especially Mignolo and Walsh, 2018) does not hone in on the structuring role of wantonness enough, to my mind, although it can sympathetically be found at almost every turn of the tale the authors tell. It is true that coloniality involves an ideological disengagement with conflicting worlds. This is what allows coloniality to present single worlds as the only world. It is what accounts for its quality of "totalization" of worlds. But the *moral* condition of this disengagement is a failure to understand autonomy in community (see Chapter 2 of this essay), and this is at bottom a failure of moral accountability to others and to the world as it emerges in disagreement between worlds. The failure of *moral accountability* at the heart of coloniality is what makes possible its totalization of worlds, it's false and *a priori* universality (that is, a universality worked out before *working through* to shared worlds through the disagreement of those who reject the given world). This failure of accountability is another name for selfishness, and its result is wantonness. For when, in the spirit of coloniality, societies and their people spread out acquisitively and instrumentally, merely practically and not relationally, their lack of moral accountability leads inevitably to wanton behavior. They harm others and disrupt their worlds without account, seeking selfish aims. Thus, we should be speaking of the *core moral problem of coloniality*, its wantonness predicated off of a morally arbitrary ethos, even before we focus on "totalization."

132 See the realization at the conclusion of Chapter 3 of this novel. Love may seem an odd thing to invoke. But consider Kant's understanding of *moral* love, which translates a religious understanding of good will and responsible care for one's fellow person. See Immanuel Kant, *Religion Within the Limits of Reason Alone*, translated by Theodore M. Greene and Hoyt H. Hudson, New York: Harper Torchbooks, 1960. Of course, Kant was to some extent caught in the "colonial matrix of power" (Mignolo, 2011). He rejected relational reasoning in his suspicion of sentimentalist morals. Rousseau, by contrast, fared better, in large part because he philosophized relationally. Interesting, however, that Kant so admired Rousseau.

133 Cf. Coulthard, 2014; Whyte, 2018b.

134 Cf. Mignolo, 2011, Afterward, especially his running discussion of "de[W]esternizing freedom," setting freedom within a "pluriverse," not "universe," of communities and traditions with their own worlds, including their own approaches to human relationality, sense-finding, and ease of living.

135 Here, as I mentioned, I part ways with Vogel, 2015, whose entire logic of environmental, social construction is practical, consistent with the repressed coloniality of even Marx. Marx, too, was part of the problem. Marx, too, was colonial. But perhaps we can begin to restore the "social" (i.e., relational) back into "*social* construction?"

136 Another way to put this is that if good relationships are the key to decolonial work and if good relationships involve land, then I must begin unlearning a view of being a person that denies relatability and landed relations in one, violent abstraction. On decolonial unlearning see, de Jong, Icaza, and Rutazibwa, 2019.

137 Whyte, 2018a; Pasternak, 2017; Coulthard, 2014.

138 Whyte, 2018b, challenges settlers to face what they think being an "ally" really involves. Will we settlers actually stand against laws and the breaking of treatises that "continually" and "collectively" repress historical crime and broken or non-relationships with the indigenous and their worlds of land?

139 What are *relatable lands*? What does shifting our land logic do? What does it mean to say that land can no longer be merely practical, merely a condition on agency or an object of use, but must matter as an interwoven, historical inheritance across worlds? Lands convey violence in a society structured by coloniality, projecting more violence into the deep time of the Earth and *homo sapiens*'s remaining history on this planet.

Here is "humanist adaptation" not considered in Thompson and my 2012a, section 1 on ecological restoration. Just so, *our* volume was part of the problem! Here, instead, is a form of *meta-restoration*, if you will, a restoration of the *terms* of the land as a field in which one understands land as potentially restorable at all, and *what that would even mean*. Singer (2000), Gardiner (2011), Thompson and Bendik-Keymer (2012) were all written in coloniality.

140 Social Justice Institute Leadership Collective, "Land Acknowledgement, March 8, 2019," via Susan Dominguez, personal correspondence, March 8, 2019. The full text reads:

> In recognizing the land upon which we reside, we express our gratitude and appreciation to those who lived and worked here before us; those whose stewardship and resilient spirit makes our residence possible on this traditional homeland of the Lenape (Delaware), Shawnee, Wyandot Miami, Ottawa, Potawatomi, and other Great Lakes tribes (Chippewa, Kickapoo, Wea, Piankashaw, and Kaskaskia). We also acknowledge the thousands of Native Americans who now call Northeast Ohio home. Case Western Reserve University and the Greater Cleveland area occupies land officially ceded by 1100 chiefs and warriors signing the Treaty of Greenville in 1795.

141 Cf. my 2016, where I first discussed an "anthroponomic orientation."

142 Cf. my 2018e on the "anthroponomy criterion" and the multiple names of the "Anthropocene." See also my 2019.

143 Cf. my 2006, lecture 2 on mountain top removal in Irian Jaya, Papa New Guinea in the traditional land of the Amungme; and also Whyte, 2018a.

144 Rachel Riederer, "The Other Kind of Climate Denialism," *The New Yorker*, March 9, 2019, accessed September 29, 2019.

145 Steffen et al., 2018.

146 See Chapter 1 of this novel.

147 Cf. Pasternak, 2017.

148 Whyte, 2018a.

149 I think here of Bruce Kafer's Lakota teaching before the upper school of Hathaway Brown, January 11, 2019, as part of the program, "A Climate Change on Climate Change." Bruce spoke of the "great mystery" that anchors Lakota cosmology.

150 The Population Reference Bureau (PRB) estimates that 108,000,000,000 (one hundred and eight billion on the short scale) *homo sapiens* have lived in the last 50,000 years. PRB, "How Many People Have Ever Lived on Earth?" accessed March 14, 2019. Obviously, this figure is a figure of speech.

Figure 5.1 Writing desk, winter 2019, Shaker Heights, Ohio – the once kitchen table of my grandmother from Pleasant City, Ohio, circa 1930s

5 How should I involve anthroponomy in the course and prospect of my life?

In the sun of the study, Shaker Heights, Ohio, once land of many nations, taken by broken, United States treatises, early spring

When a word is broken, how do we use it while taking in the wrong that broke it and, by facing what is wrong, make sense between people at least possible again from out of the senselessness of the wrong?[1] One way is to ground broken words in accountability. Good relationships depend on – and begin with – accountability to each other.[2] Only in accountability is trustworthiness.[3]

But the world of broken words is an arbitrary flow of half-sense and of non-sense.[4] Accountability freezes a broken world's swirling mess and cracks so-called "common" sense open to the moral avoidance beneath it. Only then can the splinters of words become connected with the wrong of the broken world such that the sense of the relationship appears first as demanding restitution. The wrong must be made right. The word was given, wasn't kept out of the wrong in the rela-tion, and now, facing that, it is time to make things right. It is always time – high time, we might say, as an overtone to daily life.

Perhaps it would be easier to rid one's world of a word that is fallacious and cast it to the void. It seems easier to start over. But words broken by wrongs make up a wrong world. In escaping it, we avoid autonomy in that world, an autonomy achieved only by working through disagreement. We end up avoiding good relationships.[5] This is especially problematic when we take stock of our lives on Earth.[6] Our planetary situation of implicated ecologies and entangled political economies carries what we do such that we do it affecting each other. This brings moral accountability from near to far and back, landing on us like a ceaseless ocean of waves.[7] But if the world in which I've grown to the middle of my life is structured by wrong,[8] to avoid it – including its structuring words – is to avoid accountability brought to me on those waves.

Morally speaking, we have to square up[9] – like it or not – with what structures this world, including its *structuring words*. We have to let the waves crash around us. The "Anthropocene" is a structuring word of the world structured by coloniality and predi-cated off of actually existing capitalism. The wrong is encapsulated by colonialism as a set of failures of moral accountability, and the broken words of colonialism – the failed treatises, hypocritical appeals to missionary "humanization," and the so-called "benevolence" of colonialists – appear in the words of colonialism's world.[10]

When I hear or read the "Anthropocene," I look at the range of uses – that is, the application and interpretations – of this word broken by the broken world it hides. I look at this word's colonial establishment. The "Anthropocene" has left its mark, just as the broken world that created it has, a mark of inconceivable risk and slow violence especially to the vulnerable of the planet.[11]

From where I write today, the "Anthropocene" will almost certainly become a time-stamp.[12] Many minds and creations catch up in the "Anthropocene." Its attractive power is great. There are people around the *word*, signs of a community's *world* – a broken world.

I have come to be in this world. I cannot avoid it and start over. I live here, and the planet moves between, beneath, and above us.

The question is how I am to live with a broken word in a broken world with those with whom I disagree. If I simply dismiss the "Anthropocene," I avoid the sense of others and so avoid working through disagreement together. But if I consider the word and, in so wondering, disagree instead, I can remain autonomous in relationship with others, neither arbitrary nor dominated. That seems to be a moral way forward, both for me and regarding others.

I disagree with the "Anthropocene's" coloniality and, with that, coloniality's hiding of colonialist systems of relations, including their capitalism and industrialism that have brought our planet to the brink of a radical state-shift.[13] To my mind, colonial systems and their relations are wrong, based pervasively on the fundamental wrong of failing to be morally accountable, and it is this wrong that structures the "Anthropocene." The wrong demands a counter-sense to the "Anthropocene."[14] The wrong demands squaring up. Only then might we orient the sense of the word "Anthropocene" back toward the possibility of good relationships.[15] Maybe then we could hear "anthroponomy" in the "Anthropocene."

To relate to the "Anthropocene," I need to get clearer about the wrong that structures it. This wrong is tricky. Since the beginning of my reflections last summer, I've seen that the "Anthropocene" reifies people's agency.[16] "Humankind" is said to be "geological agent."[17] But I realized that this is imprecise to the point of being misleading and so obscures where accountability is needed.[18] Specific social processes – for instance, those bundled together in "modernity"[19] – have become powerful enough to alter our planetary situation at least for the geologically near term of tens to hundreds of thousands of years,[20] something that is not much in geological time.[21] Even the most likely geological criterion of the start of the "Anthropocene" – the minute but detectable irradiation of the entire planet's surface due to nuclear explosions and the rapid increase in CO_2 emissions on a planetary scale following 1945 – clearly points to specific social processes, such as industrial militaries and industrial, fossil fuel based, modern society following on the rise of the Cold War in a battle of production and geopolitical control between modern empires.[22] The killing of megafauna 12,000 years ago did not reach this planetary scale.[23]

How has colonialism settled our situation, structuring it? By focusing on colonialism, I'm not implying that pre-colonial human beings weren't capable in some sense of altering the planet's environment markedly. Such a claim, however, is empty and misleading without specifying that sense, a sense which is historical, constructed socially over time,[24] and which reaches a geological scale only following on the modern power of colonialism with its colonial ideology, interwoven capitalism, extractive methods, and onset of industrialism fueled by those extractive practices.[25] The "Anthropocene's" implied causal claim leaves us with a generality that obscures what we are trying to understand.

For one thing, it implies that humans are in some sense ahistorical, where history is understood culturally, not simply geologically as "natural history." Human beings weren't capable of altering the planet without constructing, over complex contingent histories, the social processes that *are* altering the planet's extant biological and geological order,[26] and the humans who came to themselves inside and through those social processes weren't just any human beings. They were self-proclaimed European colonialists, early capitalists, industrialists, and moderns most generally.[27] Those people *and more precisely the social processes that defined them continuing on with violence*[28] brought the planet to where we are now. The social processes threw the situation ahead to us and to all who come after us.[29]

By lumping indigenous forest burning, megafauna hunting, the development of agriculture, and the rise of the first human cities together with the atomic, geopolitical domination of the planet in a struggle between industrial, fossil-fuel-based economies, one hides colonialism, actually existing capitalism, and industrialism as the distinctive social processes that they are and which have determined not only a world but the planet's history.[30] This makes a specific kind of society innocent by being lumped together with all the rest.

The abstraction of the scientists seems to be itself a form of coloniality. Through natural science disconnected from social critique, the wrong spreads out.[31] Yet the displacement isn't intentional. It seems to reflect the "vicious sedimentation" of colonial cultures that remain ignorant of how completely absurd and offensive it is to think that forms of "collective continuance" that begin in ecologically reflexive land relations would inevitably produce our planetary situation (!), predicated as it is today off of a vicious form of capitalism, especially, that begins with turning the land into a calculable resource and object for exploitation.[32]

This is not to claim that indigenous societies were ecologically pure or perfect. The megafauna extinctions happened. Some societies collapsed. The point is rather that for societies that involved and that do involve ecological reflexivity[33] as constitutive of pervasive social processes and forms of governance and which have not operated in the abstractions of capitalism and of modernity more generally at a global scale or with the view of the Earth as a resource to be plundered, the societies are simply not structured by the forms of power that have led us to our planetary situation and which are driving it with slow violence and massive risk into the future.[34]

Lumping all ways of life together with the social drivers of our planetary predicament is symptomatic of colonial hiding. It structures the ongoing reproduction of colonial relations.[35] The "Anthropocene" as a categorical name of natural science is a form of colonial hiding. That is a subtle part of its wrong.

The violence of colonization and its cover-running ideology of coloniality have inaugurated the processes that are causing our planetary situation. People act for purposes they imagine, envision, or presume, and these purposes have justifications,[36] entire schemes of sense that make a world and are reflective of the world in which people carry on in the instituted practices of their lives.[37] In colonization, the globe became the spoils of the colonizer. This involved and involves ornate rationalizations that displaced and displace ecological reflexivity as a core component of being civilized. One moved and moves on to new resources in the face of used up land. One separated and separates people from their lands, capturing people into use, abuse, killing them off, or letting them waste away.[38]

The violence rationalized by colonization was and is immense. The colonizers took and take what they could and can get away with, moving fast and avoiding the very trace of moral relations with all their brawn, technology, and rationalizations, including of science. In their wake and on their ongoing basis, the relations of modern production took form, driven by early capitalism and imperial conquest, rationalizing a world that became increasingly devoid of moral relations in lieu of capitalistic economic or mass-controlled resource production.[39] It is this world that lacks a sustainable basis flying headlong into the planet's future propelled by the past's systems.[40]

Still – and here the wrong that is tricky becomes subtle – there is something strange in the "Anthropocene," an underside that works against its hiding, should one only twist it toward the light, where it becomes reflective, visible like a piece of glass flashing in the sun far off in the distance.[41] Wishful thinking is involved in the "Anthropocene," too, and the wish is for us to be collectively agential enough to shift the planet's geology. All of humankind hasn't caused our planetary situation, but the word makes it seem that all of humankind *could*. Moreover, discussing agency consistently, the agency is thus implied to be intentional.[42] The wish of the "Anthropocene" involves intentionality, too.[43]

Does the "Anthropocene" *un*intentionally suggest a political responsibility that could be collective, critical, and intentional? That would be to include all of humankind in the construction of social processes that do not depend on broken words and bad relationships, but that respect our intertwined autonomy in this planetary situation. One of the reasons I have turned to anthroponomy is that I am trying to capture this wishful undercurrent of the "Anthropocene" in a way that expresses moral accountability for a past and for systems of relations that have precipitated our planetary situation.[44] I am trying to make anthroponomy the criterion of the just invocation of *anthrōpos* in the "Anthropocene."

Early morning, Sunday, tucked in a worn chair near Mom's old piano, Shaker Heights, Ohio, once land of many nations

> The "Anthropocene" is the future. Yet given its formation, it appears as colonialism's wake. Moral accountability demands a strange reversal. Forward is backward. Looking backward, what is the way forward?
> Change the system of relations.
>
> —— Early-morning note in the back of a book on queer phenomena, after reading Michael Yellow Bird, the time between winter and spring

Perhaps we could use a critical attitude when speaking of the "Anthropocene?"[45] In the spirit of a "decolonial option,"[46] I wonder whether the "Anthropocene" might be counter-sensed?[47] Thus, I could disagree in relationship with those who use the "Anthropocene" in such an ongoing colonial way. This would involve me reorienting the sense of the "Anthropocene" by claiming a path that is autonomous.[48] What if to engage the "Anthropocene" weren't to accept vicious sedimentation and reproduce coloniality, but to resist our situation *anthroponomously*? What if anthroponomy could help our planetary situation swerve from its course while decolonizing on the basis of decolonial work?[49] My way to relate to the "Anthropocene" would then be through an anthroponomic orientation.

Suppose that the "Anthropocene" became a name not for what humankind has caused, but for the responsibilities of humankind? Being anthroponomous, these responsibilities would be inclusive, morally accountable, and respectful of people's autonomy.[50] As coordinated uses of autonomy fit to our planetary situation, they would be, moreover, planetary-minded, ecologically reflexive, and decolonial.[51] The "Anthropocene" would then name a kind of *de*volution, rolling worlds back down to earth away from abstract coloniality and toward disagreement in relationship.[52] It would name a social-ecological process of lands to be decolonized and future communities deserving our ecological reflexivity.

The "Anthropocene" would also become a name for a kind of *re*volution. It would name the task of socially constructing planetary-scaled, ecological reflexivity that isn't colonial, global management but is instead morally accountable through and in a pluriverse of relations, forms, and names of "collective continuance."[53] This complex movement – away from coloniality back down to earth; yet open to the many ways people might find a universe while squaring up with moral accountability across ecologies in planetary-mindedness – would be like *tenchinage* ("sky-earth throw") in *aikido*,[54] receiving the colonialism of the "Anthropocene" and moving with a counter-sense that redirects our social processes through their wrong and toward moral accountability in relationship.

If this were right, I'd be able to construct a complex, acceptable, relation to the "Anthropocene." One thing implicit in this thought it that the "Anthropocene" come to signify a *prospective responsibility*, not an implied causal claim.[55] The "Anthropocene" would come to mean what people in an implicated, entangled, social-ecological planetary situation should responsibly consider in the living of their lives and the reproduction or reconstruction of their institutions. To talk about the "Anthropocene" would then point to how people should be in a planetary situation such as ours when moral accountability to people's autonomy-in-relationship is taken as basic to any acceptable institution. It would be the name for what "humankind" produces and reproduces in taking autonomy-in-relationship as a moral condition facing our planetary situation. Anthroponomy would be the name of the process of squaring up with our planetary situation.

Call this a *generative notion of the Anthropocene*. It seems close to how I should relate to the broken word I've inherited with its broken world behind it. What does this generative notion demand? What should be involved in it?

Tuesday morning, coffee shop, Little Italy, Cleveland, once land of many nations, home of the "Columbus Day Parade"

One thing clear is that the Anthropocene should involve self-determination, or it will be part of the problem of coloniality shadowing actually existing capitalism and industrialism on a planetary scale. In my land, then, it will also be part of ongoing colonization – the domination of indigenous nations by arbitrarily refusing to honor treaties, erasing indigenous practices,[56] and overwriting indigenous autonomy with oppressive norms.[57] The wrong to be faced *here* is a wrong of domination, violating people's autonomy. Yet without involving self-determination, the social processes of the Anthropocene will curtail people's autonomy in many ways across the planet and into the far future, forcing them to live in a world in which they are to accept things that make little to no sense to them.[58]

The wrong of denying people self-determination is profound and basic. Self-determination is basic to moral accountability and moral equality. The root of moral judgment is autonomy, each person finding what ought to be, what makes sense to them. The basis of interpersonal accountability is the fact of people having their own moral judgment.[59] When self-determination is hampered, autonomy becomes weakened, moral judgment becomes restricted, and thus accountability to each other is undermined. Others become more and more like objects than like persons, living within that double consciousness.[60] And so an especially deep and damaging form of slow violence begins in which people are half-people, living half-lives, losing their humanity behind cycles of domination, exploitation, abuse, self-damage, and nihilism.[61]

To escape being complicit in a diffuse yet targeted,[62] modern environment of domination,[63] self-determination should shape a counter-sense structure for the Anthropocene[64] – one that swerves to align uses of the "Anthropocene" with the generation of autonomy and the decolonial opening of worlds-in-relation out of moral accountability to our planetary situation that I call "anthroponomy." Let the "Anthropocene" become a signal: Involve self-determination in our planetary situation.

This is required by anthroponomy in a qualified way. Anthroponomy includes a complex relationship with self-determination, at bottom unconditionally supportive of it provided that it rests within the moral accountability needed for good relationships. On this view grounded in moral relations, there can be no anthroponomy without self-determination and no self-determination without autonomy-in-relationship.

The question, then, is how to understand the Anthropocene as a prospective responsibility, not a retrospective causal claim. The prospective responsibility must be structured by social processes of self-determination understood as autonomy-in-relationship. At the heart of this responsibility is the priority of disagreement as a ground of trustworthy relationships formed around the mystery of the world.[65] Disagreement thus becomes crucial to the Anthropocene, held in light of our planetary situation and its immoral and unjust dynamics inside coloniality.[66] Disagreement and self-determination go together dynamically.

Seeing the issue of our planetary situation through disagreement-in-relationship and self-determination is crucial for decolonizing my land. If there is a place where self-determination *and* disagreement matter it is in confronting ongoing colonization, not simply residual coloniality.[67] Here, the issue of creating a "pluriverse"[68] is in no way mostly an issue of mentality or even of language, although it includes both.[69] The issue is one of reclaiming sovereignty within indigenous communities in and with indigenous land.[70] Here, the sense of autonomy isn't ever satisfied with something mental (if it ever is in other contexts). The issue is embodied, landed, power-structured, instituted – a reality of slow violence and ongoing, normalized domination.[71] Since the broken words of colonialism's violation of violent treaties double-down on domination, a meaningful self-determination in decolonization seems to demand land relations restituted down to the roots of land sovereignty. Colonial land sovereignty must change. This must be so, then, on account of autonomy-in-relationship, anthroponomy, *and* now the counter-sensed Anthropocene.

Justice is unappeased in my land, an overtone to daily life. It points to what I called the "land abstraction" structuring everything,[72] the stamp of the practical erasing the relational through the practices, institutions, and norms of colonialism, actually existing capitalism, and industrialism. These, in turn, situate the injustice here as part of the planetary process called, itself colonially, the "Anthropocene." Understanding my relation to the Anthropocene[73] in light of the need to protect and to promote self-determination and its restoration here in a process that has ties to planetary wantonness cannot be something I solely determine sitting in a study and writing, dictating from my mind through my fingers to the electrical page. This is not simply because the problem is so massive. The issue is more conceptual than that. When self-determination is a core concern of moral accountability, the issue cannot be for me – especially as a colonialist – to speak for what others need. Instead, out of the spirit of autonomy-in-relationship, the issue would seem to be to clear the way for what truly makes sense to people as moral equals, here, out of a history of broken words. What would anthroponomy look like with that structuring it?

Writing day, Thursday, at a table stacked with books in a colonial nation state driven by capitalism, morning

Let me back up first. Our planetary situation threatens the autonomy of people in profound ways, increasingly in the present and far into the future, including the widespread sense that the more than human world of life is not something that one just lays waste.[74] Looking at what are the negligent, greedy, heartless and unjust effects of modern social processes, including the institutional arrangements through which the processes are able to be wanton,[75] my approach has been to think about a social idea – both moral and political – in which we would subject our current social processes to a critique: Are they consistent with, not undermining, or supportive of the autonomy of all human beings? Do they avoid dominating anyone and support the autonomy of people over far-reaching time and around the planet?

This approach led me quickly to engage with the widest systems of domination, the same ones, moreover, that have structured and produced our planetary situation in an abstract and non-ecologically reflexive relationship to the land: colonialism, capitalism, and industrialism. Decoloniality then became an imperative moment of any possible form of anthroponomy, focused as it is on devolving the abstraction of the modern world's land logic. Whatever anthroponomy is, it must involve decoloniality. So I found myself a citizen of a colonial nation with "vicious sedimentation"[76] structuring my relationship to my social-ecological situation.[77] How could I relate decolonization to our planetary situation?

I did not even know how to name it,[78] that is, my situation, looking backward, responding forward,[79] between self-determination and planetary concerns. I had to get clear on this name and what it implied, the structuring notion of our time. With some work, I recently found a sense of the "Anthropocene" that I can use without scare quotes: I must involve anthroponomy in our planetary situation as a social-ecological process, thereby involving anthroponomy in the Anthropocene. This general approach of the name's countersense employs a prospective form of decoloniality, demanding decolonization through its commitments to respect self-determination and to address broken words.

I now need to understand what anthroponomy more exactly means for my life in that process so that I can find an orientation to my everyday action and engagement beyond the wide and deep malaise with which I began my reflections last summer. I know this: Autonomy-in-relationship rejects colonial domination. My moral and ethical assumptions concerning anthroponomy are also relevant here.[80] On the one hand, there is my basic moral commitment to respect people's autonomy, where autonomy is understood as relational around the mystery of the world. On the other hand, the Anthropocene is a situation calling for ecological reflexivity on a planetary scale. The planetary form of the threat to autonomy is what takes us, I think, to anthroponomy as a specific use of autonomy to coordinate a morally accountable response to our planetary situation, beginning with opposition to domination.

The Anthropocene involves a colonial effect – one that will be with all of humankind most likely for the rest of human time.[81] So while the question of how

I should relate to the Anthropocene passes through anthroponomy as a decolonial option, it involves decolonization with a scalar dimension that is planetary. This, however, is complicated, for it seems to reproduce coloniality's global fixation.[82] How can self-determination be both local and planetary-minded at once, and how is such a question not a form of coloniality once again? Grasping anthroponomy demands that I understand how decolonization and the wider reaches of decoloniality can cohere with problems of a planetary scale such as those involved in the Anthropocene. Can they, and if so, how?

The moral wantonness of colonialism courses through a merely practical approach to lands, predicated off of manipulative, dishonest words and the failure of moral accountability to others. Self-determination, justly resisting colonialism, marks the failure of moral accountability, the mere objectification of land torn up wantonly and violently from beneath people, and the false words that have tricked and exploited agreements that were never really done in good faith. Beyond self-determination, good relationships depend on relations through lands, finding autonomy-in-relationship with others even on a planetary scale. Self-determination is needed, then, because autonomy-in-relationship and relational reasoning between people and through lands are erased, undermined, and destroyed by the logic of coloniality and the practice of colonialism.

The planetary situation into which colonialism and its emergent extractive and industrial processes have flung all of us places an additional consideration on self-determination, relationality, and lands. Self-determination today will occur in the context of planetary change affecting the far future. Relationships will occur across vast reaches of the planet's space and time, and lands circulate through climactic and other planetary processes that entangle them. Due to the planetary reach of what we do with our lands in our planetary situation, autonomy thereby is allochthonous.

To say that something is "allochthonous" is to say that it involves what comes from elsewhere, rather than what is indigenous.[83] The etymology means what is of "other earth."[84] In the Anthropocene, wherever we live is always implicated in "other earth" – in the planet's circulating and long-term conditions as driven by processes that emerge around the globe. Exactly in this way, the question of autonomy-in-relationship coursing through lands becomes planetary in the Anthropocene. The ironic result of colonialism – the inaugurating, allochthonous process of modern society[85] – is that the self-determination that opposes colonialism must, in the Anthropocene, demand a freeing of social processes everywhere to respect the autonomy-in-relationship of people in our planetary situation.

I can put it like this: Respect for and the restoration of self-determination in the Anthropocene should involve an *allochthony condition*. According to this condition, the self-determination of a community structured by a form of autonomy-in-relationship is to be respected insofar as that community's relationship to the land does not continue to violate the self-determination of *other* communities by being wanton with or laying waste to the land these communities share across space or over time.

What is ironic about the allochthony condition is that it returns the right to the land to indigenous societies practicing land stewardship and seeking a non-extractive, sustainable economy.[86] It deprives colonial nations, as structured by coloniality, of their moral right to the land, since they are predicated off of a merely abstract, practical relation to land complicit with wanton resource use and the extractive driving of the planetary system.[87] The inaugurating allochthony of colonialism thus boomerangs back to deprive colonial nation states of the moral legitimacy not only of their already criminal sovereignty but *also* of their land management processes and institutions.

That is a surprising turn of sense. Self-determination in the Anthropocene doesn't imply simply a right against colonialism. *It implies a right to reform the entirety of land systems wantonly used in colonialism and erased as relational fields by coloniality's reduction of them to practical objects.*

Due to the planetary condition of the Anthropocene, self-determination becomes far more reaching than even taking back colonized or treatise-violated land. It involves a claim to land that is driven wantonly to undermine the autonomy-in-relationship of others, especially into the future. This claim implies the right to demand a change in the way land is managed, beginning with the demand that it be viewed relationally in light of the autonomy of people far into the future who will inherit in and have a right not to have to bear the burden of wantonness.[88]

The idea here is simple, although the realization of it is, of course, complex: You lay waste to it, you lose it. The land isn't anyone's property as a practical object. It is a living field in which we can live up to our moral responsibilities or fail to. Through the land, countless generations of people and other living beings lived, live and will live. Do the social processes that shape the land do so in ways that respect the autonomy-in-relationship of others, especially into the future? Insofar as the more than human world of life is a significant part of people's relationships *at the least* by virtue of its moral considerability,[89] do the social processes that shape a given land also eschew the disruption in sense of disrupting those relationships? If the social processes aren't respectful, why would they have a moral right to continue as they are? They violate autonomy-in-relationship. They are *wanton.* They betray a lack of relational reasoning and a failure of good relationships.

In the Anthropocene, decolonization demands more than an honoring of treatises. It demands that societies structured by coloniality prove that they can become worthy of the land. Otherwise they should lose control over it.[90] That is, people have every right to reconfigure their processes and construct something that could, collectively, continue in a moral way.[91]

Call this relation to the Anthropocene a kind of *landed autonomy-in-relationship*, where the land is planetary in its implications and entanglements. The land belongs to no one. One belongs to the land. The question is what this means.

Sunday evening, in Misty's yellow chair in the basement, the spring rains outside in this colonial nation state

I want to get clear about lands at this point, since they have become central to the entire torque of this essay's course over the last half year. How are they involved in autonomy-in-relationship? What does it mean to say that a society is "eco-logically reflexive" with them?[92] I need to get clear on these questions so that I can understand how to work to clear room for self-determination that is mor-ally accountable within our planetary situation and stretching through planetary scales. This work will be decolonization here and now in my land, yet also in a sense fitting the Anthropocene. Now involving anthroponomy, the Anthropo-cene involves our prospective responsibility to co-construct social processes of autonomy-in-relationship accountable regarding issues of planetary scale.

As I've already substantially considered,[93] the "land" is shorthand for a com-plex social-ecological *response* to the mystery of the world of life and the seeming infinity of the cosmos that, most precisely, becomes our human environment.[94] It is not some preexisting moral system that makes claims on us prior to our relation-ships with it. What any ecological system in which we live is, however, is a pos-sible source of relationships, and this is so even in industrial systems, mechanical environments, and so on.[95] Moreover, in any ecological system in which we live, our practices slowly construct, or better *re*construct, the environment, drawing on relationships and making them possible, responding to the "outside" of the system we think we have made and that we face in wonder or awe.[96] Usually, this recon-struction is habitual or in a limited way intentional. Always, it rests on an underly-ing "groundlessness" that is the unknown and excessive outside of the world we have not comprehended and cannot master but which we at best evoke in poetry, spirituality, and the like.[97] The practices and relationships sediment over time so that we inherit them as a pre-configuration of our situation, a place we come into and a way of being in our world.[98] They become our given environment. To speak of the "land" in this theoretical context is to speak of a specific constructed *and* responsive way of configuring *and* invoking our environment.

The land is a field of life that, with work, can sustain us and to which we relate as part of our meaning-finding and making throughout culture and our lives.[99] In the land, the more than human world of life can flourish, for a time, in varying degrees and ways. The land as a moral concept also presumes our commonly human, socialized view that life is not something one just lays waste.[100] A view to the contrary to this one is a vice and is, for instance, a result of capitalistic or colonial practices. Yes, this means that in a very real way, colonial worlds do not have lands. They are "unlanded," abstract and violent.[101]

We are dependent on lands, but only by integrating with them.[102] What this means is that we have to make a home in lands. We must develop practices and relationships predicated off of thoughtfulness with life in which we can also flour-ish.[103] In this context, some form of ecological reflexivity is already implicit in our relationship with lands. To be thoughtful with the more than human world of lands and to flourish in them, we must gain feedback from them about how life fares and the prospect of our flourishing – at the least of our capability to do so. We must also reflect on that feedback and, aiming to be thoughtful and to sustain our capacity to flourish, at times reevaluate how we consider our land. That is the most authentic point of reflexivity, and, through it, our integration with lands must be open and dynamic.[104]

All of this goes into the cultural formation of "lands" in the sense I am using the word. Obviously, there are more or less degrees and many different kinds of ecological reflexivity and thoughtfulness that can be developed consistently with lands. That is why the term is general. It designates a whole family of environments people construct, all of them predicated off of an underlying commitment to be thoughtful with the more than human world, morally accountable, and ecologically reflexive. The point about colonialism – especially in the actually existing capitalism that charges through colonial preconditions into globalized wantonness – is that the relation to the land is highly minimized if not eradicated, and the forms of that relation are themselves often superficial, shortsighted, unjust, or ultimately undermining of people's capabilities and of the more than human world. They cast life to the void in a wasteful and wanton violence, even if unintentionally.[105]

Lands, in other words, are environments predisposed to moral accountability and to autonomy-in-relationship, because they presume thoughtfulness. They are already moralized. That is how they have been socially formed, using relational reason and associated ethical[106] practices weathered into place to support the capability of people to seek a life of sense.[107]

What does it mean to say that a land involves relational reason? This has been a question since I first came to relational reasoning several months ago when considering how good relationships are formed.[108] Relational reason exists in a *space of personification*.[109] When we reason relationally, we use our receptive and imaginative capacity to understand analogically and feelingly – through emotional intuitions – how others are, personally, as well as how the more than human world can be related to personally.[110]

This is personification in a subtle and non-simplistic sense. It is not simply slapping a face onto something faceless and thereby being deluded. The space of personification begins with the notion of the difference between myself and others. Only on the basis of this difference can there be relationship. Moreover, the difference may go so deep as to show up ways in which I relate to another being that is more than a person, that is, in this precise sense, impersonal. The logic of the personal involves the logic of the impersonal. Thus when it comes to personification, the precise use of it will involve finding relationships across a range of differences, including ways in which a being is not a *person*, while one still can, in various qualified and specific ways relate to that being *personally*.[111] That is how things work when one relates to the land – or, more precisely, when one relates in a variety of ways to the many beings of one's land.

The thing about relational reason is that it moralizes relationships.[112] It brings one into the space of personal accountability.[113] If I relate to something personally, even if it is not a person, I hold myself – or am ready to hold myself – accountable to it. The space of personification extends the notion of justice – which begins between people – to the impersonal world in qualified and specific ways. For instance, if I relate personally to the flora and fauna of where I live, I consider wantonness toward them to be a kind of wrong that has failed to be accountable to the ways in which the flora and fauna have a life of their own. It is on the basis of this extension of justice that it is not absurd for people to be trustees of the more than human world that cannot speak in a court of law and to imagine a form of human justice in which trustees of the land advocate for it on the manner of advocating for a being worthy of justice.[114] It is in this broad way that something scholars have called "ecological" justice may begin to make sense.[115] When a land is moralized, one must hold oneself personally accountable to it, and this may involve a complex bundle of relationships and practices to do so, distinguishing within one's land between many kinds of beings with their own life and considering how it makes sense, being thoughtful with one's land, for you all to live here together.[116]

The point is, when people are thoughtful with their lands, they must hold themselves accountable to them, and this means, to the life that goes on within them. Obviously, what this involves for any given land is highly complex and demands disagreement. The sense of the world is not straightforward in the case where we personify in the presence of what is more than personal. But the details of this are not the main point for my thinking about the Anthropocene. The main point is that when lands are moralized, they are not mere resources. The self-determination of anyone on them cannot be merely practical. The land cannot be fodder for our profit or our ego.

The Anthropocene, as the prospective responsibility to involve anthroponomy in the social processes producing our planetary situation, demands what scholars call "ecological reflexivity" in our relationship with lands.[117] Such reflexivity involves reassessing our institutions, values, and who we take ourselves to be (our self-conception) so that we remain thoughtful with the land while "thinking like a planet."[118] It is a kind of moral authenticity,[119] one that is scaled to the planetary situation and aimed at being accountable to autonomy-in-relationship across time through the relation of lands that will be respected or wasted for future generations of people and the myriad forms of life depending on our accountable thoughtfulness to them. In the Anthropocene, good relationships begin with such planetary-minded thoughtfulness with lands.

Monday afternoon leading toward evening, the spring outside in full push of life, in this nation where self-determination is undeserved given the wanton relation to the land and to the planet

Although it is still very general, I now have a clearer sense about how I should relate to the Anthropocene. The conclusion is surprising and sets out a puzzling and compelling challenge for how I should finally clarify what my moral commitment to autonomy in our planetary situation should demand of me. I should relate to the Anthropocene as a prospective responsibility to decolonize colonial states so that the self-determination and prior agreements made with colonized people are respected. Moreover, I should refuse any social process a moral right to the land when it remains thoughtless with it, that is, when, precisely speaking, it sees the environment as its practical resource and not as a land in the sense I have developed it through good relationships. This latter conclusion applies especially to capitalism and industrialism.

These are radical, decolonial conclusions – so utopian as to be disturbingly impractical. But this world as we know it is in need of change and stands on the brink of a void.[120] And – no matter – autonomy always demands seeking sense. *Sense* has led me to such conclusions, even if they remain critically ideal like the regulative ideas of Kant![121] I had no intimation of these conclusions many months ago when I began my anxious reflections. Perhaps this is the point is essaying thought – to come upon something, to you, that is novel, self-determined, (closer to) autonomous.

In any event, the conclusions I've reached go to the root of coloniality, do not deny settler colonialism,[122] and intensify decolonial claims to deny the right of nations – and of social processes more generally – that are thoughtless with the land. They imply a moral revolution within my world. *No nation has a right to the land if it isn't thoughtful with it.* Societies whose social processes are not thoughtful with the land violate autonomy-in-relationship with people across time and around the planet, including people who are trustees of the more than human world, minding it thoughtfully through good relationships. In the Anthropocene, thoughtfulness demands a planetary reach. What we do with our land – or what we do in failing to relate to the land but in using it merely as a resource, practically – affects the entire planet in indirect and aggregative ways.[123] To be morally accountable to others, societies – beginning with their communities and the people who make them – must develop good relationships through the land, "thinking like a planet."[124]

One question, of course, is how this general moral outlook coheres with prior claims of self-determination as found in decolonization claims. The answer, I think, is simple, even if in practice the answer is exceedingly demanding. People have a right to be autonomous. It is a basic moral right. Self-determination that is morally accountable to others is involved in that right to autonomy. Moreover, in cases where the colonized claim self-determination against a history of domination and broken words, there is special reason to honor those claims and make good on broken words. A history of domination is a violation of autonomy, and going back on agreements under almost all circumstances is a failure of autonomy-in-relationship.[125] Even more, in a surprisingly wide array of instances of ongoing colonization, the indigenous society that is colonized has a relationship with the land, in the precise sense I've been studying – a relationship that is thoughtful with the land, morally accountable, and open to wider reaches of thoughtfulness involving the planet, even if, historically, those reaches have not been an explicit focus of a regional, ecological sensibility.[126] On this third reason alone, the social processes of the indigenous have a moral right to live with the land and not the social processes of the colonizer.

*

These conclusions, strangely, leave me with calm inside my body – the first calm I have felt in months. But that is not so strange when I think about it. To live in injustice is a terrible thing. It takes its toll on everyone, whether they know it or not. In my case, it sits as insecurity in my being. I know that my society is

wanton – heedless of moral accountability, unjust, egotistical, violent, and dominating. I know that it has no right to the land it occupies. I've always sensed this
in the strange abstraction and violence that have filled the television and films
growing up and which now fill the internet, erupt monthly in wanton killings and
daily in egotistical and insecure aggression that people show to each other when
doing something as simple as driving or when working practically together. Competition is everywhere, insecurity is large, and the ghosts of my society haunt us
all the time if only I look: There! Racism, patriarchy, and the harrowing silence
and delusion of our desire to possess and to control the environment in a ring of
egotistical property that protects one from others and manifests one's dominating
status in an unequal society.[127]

It isn't so strange, really, that I should have felt overwhelmed going into this
process almost on a year ago. Why would I think that my society is trustworthy
when it does not have a relationship with the land and when that practical relationship is bound up with the historical objectification of people in historical crime?
Why would I trust a society predicated off of broken words?

To realize that I should respond to the Anthropocene by advocating for landed-
autonomy-in-relationship on a planetary scale, that is a relief. It helps me criticize
the wantonness and injustice of my society. It aims for something restorative.[128]
It provides a way.

The way is what I am calling "anthroponomy." Rather than reproduce coloniality, I should respond to the Anthropocene by involving anthroponomy in it.
I now see much better how anthroponomy involves a decolonial orientation. What
remains for me is to bring anthroponomy as my responsibility in the Anthropocene closer inside daily life. How should I involve anthroponomy in the course
and prospect of my life?

Answering this question will take me close to being able to walk outside with
my head up. Sometimes, as I have found this past year, the way forward is backward first. There, in the "backward" of colonial history, I found the root of the
problem. And now I must go forward to work on it with others as soon as I know
what more should be involved.

Notes

1 On "wrong," see Jacques Rancière, *Disagreement: Politics and Philosophy*, translated
  by Julie Rose, Minneapolis: University of Minnesota Press, 2004, chapter 2. "Sense"
  here refers to the sense of autonomy, namely, that the world makes sense *to* you. No
  one is completely autonomous if the common sense of their world does not make sense
  to them, if, that is, they must bear the sense of others when it makes no good sense to
  them. Obviously, autonomy is a matter of degree, but in the base of a moral wrong built
  into the fabric of common sense, that degree is severely restricted.
2 The fundamental wrong of relationships is to eschew accountability. Due to failed
  moral accountability, all other moral failures occur, continue, or are ignored. Mistakes,
  even, become negligence.
3 This is one way to read the outrage in Eve Tuck and K. Wayne Yang, "Decolonization
  Is Not a Metaphor," *Decolonization: Indigeneity, Education & Society*, v. I, n. 1, 2012,

pp. 1–40. Settler words, even of so-called "allies," do not translate into accountable actions. They become "innocence" devices to put off acting to relinquish stolen land or to honor broken treatises – or to thoroughly make way for indigenous self-determination. See also Kyle Powys Whyte, "Settler Colonialism, Ecology, & Environmental Injustice," *Environment & Society,* v. 9, 2018, pp. 129–144.

4 Cf. my "This Conversation Never Happened," *Tikkun,* March 26, 2018, accessed September 29, 2019.

5 See Chapter 2 of this novel for the importance of working through disagreement for continuing autonomy and trustworthy relationships.

6 "There is no planet B," a common climate justice protest sign of the 2010s exclaims. See also Peter Singer, *One World: The Ethics of Globalization,* New Haven: Yale University Press, 2000, especially "One Atmosphere." However, note my comments in Chapter 5 regarding Singer's coloniality.

7 Cf. William Connolly, *The Fragility of Things: Self-organizing Activities, Neoliberal Fantasies, and Democratic Activism,* Durham, NC: Duke University Press, 2013.

8 Walter Mignolo, *The Darker Side of Western Modernity: Global Futures, Decolonial Options,* Durham, NC: Duke University Press, 2011.

9 I.e. face what is wrong; be accountable.

10 Mignolo, 2011 and, on "benevolence," Jaskiran Dhillon, *Prairie Rising: Indigenous Youth, Decolonization, and the Politics of Intervention,* Toronto: University of Toronto Press, 2017.

11 On risk, see Will Steffen et al., "Trajectories of the Earth System in the Anthropocene," *Proceedings of the National Academy of Sciences,* v. 115, n. 33, July 2018, pp. 8252–8259. How *could* the risk of the non-linear and unknown be conceived? On violence, see Whyte, 2018, who adopts the term of art "slow violence" in various ways; and on the most vulnerable (the poor or dominated, future generations, and the more than human world), see John S. Dryzek and Jonathan Pickering, *The Politics of the Anthropocene,* New York: Oxford University Press, 2019, Chapter 4 on planetary – not "global" – justice.

12 A Google search done today, March 12, 2019, came up with 5,510,000 hits for "Anthropocene," and as of this writing it seems very likely that the term "Anthropocene" will be formally adopted to name the age begun with the irradiation of the rock layer on a planetary scale and the rapid increase of CO2 emissions around the planet from roughly 1945 onward. Naomi Oreskes, "Is Climate Change the End? And if so, the End of What?" Cleveland Humanities Festival, Case Western Reserve University, March 22, 2019.

On the notion of a stamp, cf. Reiner Schürmann, *Heidegger on Being and Acting: From Principles to Anarchy,* translated my Marie Gros, Indianapolis: Indiana University Press, 1987. In Schürmann's reading of Heidegger, the metaphysical *Prägung* (stamp) is that of "technology" – of seeing being as a form to be controlled through fashioning, e.g. when the planet is formed by *anthrōpos* (humankind) in the "Anthropocene." The stamp of the "Anthropocene," given by geologists, would then reflect an underlying metaphysical form giving shape to being.

13 See Steffen et al., 2018; Mignolo, 2011.

14 Not simply "*non*-sense."

15 Readers of Wittgenstein have spoken of his philosophy as "leading words home," that is, from the non-sense of metaphysics to ordinary contexts of meaning. Which ordinary contexts? Well, ordinary life is built around relationships. But are these relationships simply *back* there, in whatever ordinary contexts we've inherited, including colonial ones?

Wittgenstein can be read as a relational philosopher (conversation with Rupert Read, Russell Square, London, September 10, 2019). On my past reading, there is at least a relational interpretation of Wittgenstein developed by the proto-relational philosopher Stanley Cavell (of "passionate utterances," a relational spoken form next to practical speech acts of Austin and the theoretical designations of Russell).

Still, my question here is, what would it be to lead words home *decolonially*? What would it be, for instance, for settlers to relinquish their hold on sense? The ordinary contexts Wittgenstein often invokes are plausibly colonial. Accordingly, my working thesis is that leading words home *in colonialism* must involve leading words *back* (where the broken words lie) and *forward* (through disagreement) to good relationships. This may produce a more agitated Wittgensteinianism than is usual.

    Cf. Stanley Cavell, *Emerson's Transcendental Etudes*, edited by David Justin Lodge, Palo Alto: Stanford University Press, 2003, p. 204; and the place where he "led words home" most in the context of relationships (which he calls "acknowledging"), *The Claim of Reason: Wittgenstein, Skepticism, Morality, and Tragedy*, 2nd ed., New York: Oxford University Press, 1999. Richard Eldridge discusses "leading words home" as related to Wittgenstein in *Leading a Human Life: Wittgenstein, Intentionality, and Romanticism*, Chicago: University of Chicago Press, 1999, p. 118. I learned the expression repetitively and in practice through Leonard Linsky's Wittgenstein reading group at University of Chicago, 1994–1999 and also in the classes of Irad Kimhi at Yale University and University of Chicago in the 1990s.

16 See Chapter 1 of this novel.

17 Dipesh Chakrabarty, "The Climate of History: Four Theses," *Critical Inquiry*, v. 35, n. 3, 2009, pp. 197–222.

18 See Chapter 1.

19 Mignolo, 2011.

20 Steffen et al., 2018.

21 Matthew Adams, "Anthropocene Doesn't Exist, and Species of the Future Will Not Recognize It," *The Conversation*, March 11, 2019.

22 Oreskes, 2019; cf. Moishe Postone, *Time, Labor, and Social Domination: A Reinterpretation of Marx's Critical Theory*, New York: Cambridge University Press, 1998.

23 Oreskes, 2019, agrees. There is an interesting epistemic comparison to be made between the role of the criterion of the moment of planetary effect in the "Golden Spike" that stratigraphers seek and the criterion of the synching of the economy in the Information Age making it truly "globalized." Globality is defined by synchronicity on a planetary scale. Everything in the planet is together under a specific condition at a given moment. From the Nuclear Age to the Information Age, we are in globality. We might then ask whether the construction of *technologies of global effect* ought to be the criterion of geological naming. See Manuel Castells, *The Rise of the Network Society*, 2nd ed., Cambridge, MA: Blackwell, 2000.

24 Cf. Steven Vogel, *Thinking Like a Mall: Environmental Philosophy After the End of Nature*, Cambridge, MA: MIT Press, 2015.

25 I think this way of putting things would be my starting contextualization for a similar narrative by Martin Heidegger in "The Question Concerning Technology," in *The Question Concerning Technology, and Other Essays*, translated by William Lovitt, New York: Harper Torchbooks, 1977. In so being, I qualify now my concerns – which were also aimed at re-inscribing Heidegger's – from my "The Strange Un-agent of Our Species, Our Collective Drift," ISEE/IAEP Conference, Nijmegen, NL, 2011.

26 The biological and the geological are joined in the history of life on Earth, despite their often different time scales. See the overall thesis found when visiting The Museum of the Earth, accessed September 28, 2019.

27 Mignolo, 2011; Postone, 1998.

28 Whyte, 2018, re. "collective continuance."

29 It is surprising to me that prominent natural scientists do not use the precision of the social sciences just as they would expect social scientists and humanists to use natural scientific precision. I recall a particularly surprising exchange in 2016 with Carl Safina in person at Case Western Reserve University and over email on this matter. Safina refused to think that there is any benefit in seeing which social processes have produced the planetary situation we are in. Thereby, he contradicted his own practical recommendations about *better* ways of interacting with the intelligence of more than

human animals. Oreskes (2019), too, while aware of the debates of the "humanities" over the name "Anthropocene," simply sidelined them as "different" meanings, a surprisingly unrigorous response from someone trained in both natural and social science and viewed as a prominent humanist. Natural scientists, or those working with them, betray a latent prejudice against the humanities and social sciences in these instances, in my view, since they dismiss or minimize claims without (a) appreciating them or (b) reasoning against them.

30　Cf. Heather Davis and Zoe Todd, "On the Importance of a Date, or, Decolonizing the Anthropocene," *ACME*, v. 16, n. 4, pp. 761–780.

31　This disconnection shouldn't be surprising. Not only has natural science historically railed against social science as epistemically flawed, but it has allied itself with value neutrality prior to critical science studies which only arose in the late twentieth century. Moreover, modern science evolved historically alongside industrialism and capitalism and has been complicit in colonialism from the beginning with its early ideologues, such as Francis Bacon, who even viewed the Earth as a resource to be plundered. See Caroline Merchant, *The Death of Nature: Women, Ecology, and the Scientific Revolution*, New York: HarperOne, 1990; Steven Shapin, *The Scientific Life: A Moral History of a Late Modern Vocation*, Chicago: University of Chicago Press, 2010.

32　Consider this extended quotation from Waziyatawin and Michael Yellow Bird, eds., *For Indigenous Minds Only: A Decolonization Handbook*, Santa Fe, NW: School for Advanced Research Press, 2012, p. 4 (added connections mine):

> We are heading into a new era in which we will no longer be able to deny the effects of industrial "civilization's" grave damage to diverse zones and ecosystems of our planet. As Indigenous Peoples we know that this devastation has been occurring on Turtle Island [i.e., what colonizers call "North America"] for the last five centuries; and, for that length of time, our ancestors have continued to sound the alarm to the ongoing, un-restrained feeding frenzy of non-renewable resources by the corporate-led, capitalist engines of colonial society. We know from the stories and prophesies of our ancestors that this mindless [i.e., wanton] consumption activity portends a future of deep hardship and dramatic change for all life on earth. . . . [T]he planet has reached a "tipping point" [misnamed the "Anthropocene"] that will undoubtedly threaten the foundations [i.e. non-ecologically reflexive political economy] of industrial civilization and the survival of much of life. The whole earth will reap the effects of hyper-exploitation, exceeding the carrying capacity, and wide-scale ecological degradation and destruction. To be clear, this present and impending disaster was not the making of our ancestors who maintained indigenous, sustainable ways of life – ways that were considered to be backward, primitive, and underdeveloped by our colonizers [i.e., in coloniality].

See Whyte, 2018 on vicious sedimentation and collective continuance; Dryzek and Pickering, 2019 on "ecological reflexivity"; and Geoff Mann and Joel Wainwright, *Climate Leviathan: A Political Theory of our Planetary Future*, Brooklyn, NY: Verso, 2018 on capitalism's role in our situation. See also Chapters 3 and 4 of this novel.

It does not strike me that the well documented instances of indigenous, unsustainable activity – e.g., mass buffalo slaughter, the mass and wasteful killing of the mastodons (see Matthew Ridley, *The Origins of Virtue: Human Instincts and the Evolution of Cooperation*, New York: Penguin, 1998) – disprove that the collective continuance of much indigenous society strives to be ecologically reflexive in a manner unknown to industrial modernism and especially the history of capitalism. *Can we honestly say that industrial society or capitalism strive to be ecologically reflexive?*

33　Again, Dryzek and Pickering, 2019 for this important term, the new "first virtue" of social institutions. Dryzek and Pickering do not explicitly recognize this reflexivity as decolonial nor as having been consistently indigenous in many indigenous societies.

However, in email correspondence, John Dryzek did welcome this connection between the new virtue of ecological reflexivity and its other names in indigenous land relations. He recognized that in such societies it predates "Anthropocene" conditions. Personal email correspondence November 13, 2018.

34 On the significance of "forms of power" for analyzing our planetary situation, see my " 'Goodness Itself Must Change' – Anthroponomy in an Age of Socially-caused, Planetary, Environmental Change," *Ethics & Bioethics in Central Europe*, v. 6, n. 3–4, 2016, pp. 187–202.

35 One might also call it a form of moral corruption. See Stephen M. Gardiner, *A Perfect Moral Storm: The Ethical Tragedy of Climate Change*, New York: Oxford University Press, 2012, esp. chapters 9–10 and appendix 2. Of course, Gardiner's text also involves a form of colonial hiding by assuming *homo economicus*. See Chapter 4 of this novel.

36 Aristotle, *Nicomachean Ethics*, translated by Christopher Rowe, New York: Oxford University Press, 2002.

37 Vogel, 2015; also Martin Heidegger, *Being and Time*, translated by Joan Stambaugh and Dennis J. Schmidt, Albany: SUNY Press, 2010.

38 On the prevalence of these dynamics still today, see Mike Davis, *Planet of Slums*, New York: Verso, 2006.

39 Mignolo, 2011; Postone, 1998.

40 See Dryzek and Pickering, 2019 on "path dependencies."

41 On the way what hides unjust systems may simultaneously reveal them, cf. Susan Buck-Morss, *The Dialectics of Seeing: Walter Benjamin and the Arcades Project*, Cambridge, MA: MIT Press, 1989.

42 G.E.M. Anscombe, *Intention*, Cambridge, MA: Harvard University Press, 2000.

43 Given the impersonal, *un*intentional logic of coloniality driving our situation, the implied intentionally of the "Anthropocene" is especially wishful. Cf. Mann and Wainwright, 2018, esp. chapter 5; also Mignolo, 2011 on the "CMP" – the "Colonial Matrix of Power," including capitalism, racism, and patriarchy.

44 Cf. my 2016.

45 See Michel Foucault, "What Is Critique?" in James C. Schmidt, ed., *What Is Enlightenment? Eighteenth Century Answers and Twentieth Century Questions*, Berkeley: University of California Press, 1996, pp. 382–398. Cf. Søren Kierkegaard, "The Concept of Irony, with Continual Reference to Socrates," in Hong V. Howard and Edna H. Hong, trans., *Kierkegaard's Writings*, v. 2, Princeton: Princeton University Press, 1992.

46 Mignolo, 2011.

47 As Sarah Ahmed notes, this would be to make the word *swerve* – from *quer*, swerve. See Sara Ahmed, *Queer Phenomenology: Orientations, Objects, Others*, Durham, NC: Duke University Press, 2006. See also Lynne Huffer, *Mad for Foucault: Rethinking the Foundations of Queer Theory*, New York: Columbia University Press, 2010 and *Foucault's Strange Eros*, New York: Columbia University Press, 2020, where the swerve of words is to the mystery of the world.

48 Cf. Ahmed, 2006.

49 See Steffen et al., 2018, on the courses of the Earth system.

50 See Chapter 2 of this study.

51 See Paul D. Hirsch and Bryan G. Norton, "Thinking Like a Planet," in Allen Thompson and Jeremy Bendik-Keymer, eds., *Ethical Adaptation to Climate Change: Human Virtues of the Future*, Cambridge, MA: MIT Press, 2012, chapter 16; Dryzek and Pickering, 2019, chapter 3; and Walter D. Mignolo and Catherine E. Walsh, *On Decoloniality: Concepts, Analytics, Praxis*, Durham, NC: Duke University Press, 2018; cf. Whyte, 2018 for a decolonial assertion of "ecological reflexivity" in Anishinaabeg terms, although not planetary-scaled.

52 "Devolve" from *de-* "down" and *volvere* – "to roll," *Oxford American Dictionary*, Apple Inc., 2005–2017.

53 Whyte, 2018.

54 See Sensei Susumo Chino, *Aikido: Tenchi-Nage*, Tokyo (Yoshinkan Aikido Hombu Dojo): Empty Mind Films, 2015.

55 On prospective responsibility, see Iris Marion Young, *Responsibility for Justice*, New York: Oxford University Press, 2010; see also Allen Thompson, "The Virtue of Responsibility for the Global Climate," in Allen Thompson and Jeremy Bendik-Keymer, eds., *Ethical Adaptation to Climate Change: Human Virtues of the Future*, Cambridge, MA: MIT Press, 2012, chapter 10.

56 Whyte, 2018.

57 Sarah Deer, *The Beginning and End of Rape: Confronting Sexual Violence in Native America*, Minneapolis: University of Minnesota Press, 2015.

58 Recall Susan Neiman's criterion of evil as the social production of a senseless world in her *Evil in Modern Thought: An Alternative History of Philosophy*, Princeton: Princeton University Press, 2002.

59 On moral judgment, see even a mainstream text for undergraduates: Richard Burnor and Yvonne Raley, *Ethical Choices: An Introduction to Moral Philosophy with Cases*, New York: Oxford University Press, 2018, chapters 3 and 4; and on accountability, see Stephen Darwall, *The Second Person Standpoint: Morality, Respect, and Accountability*, Cambridge, MA: Harvard University Press, 2006.

60 See W.E.B. DuBois, *The Souls of Black Folk*, Project Gutenberg, 2008 as well as Franz Fanon, *Black Skin, White Masks*, translated by Richard Philcox, New York: Grove Press, 2008. Cf. Glen Coulthard, *Red Skin, White Masks: Rejecting the Colonial Politics of Recognition*, Minneapolis: University of Minnesota Press, 2014. In my view the *specific* double-consciousness of being "whited out" while "Black" is a form of the general capacity of practical reason to subvert relational reason, e.g., turning people into objects which are subtly permitted enough specific (i.e., relational) personifications to allow use and abuse in daily life while fundamentally denying the person's capacity for self-determination, autonomy, and moral judgment.

61 See Dhillon, 2017; Waziyatawin and Yellow Bird, eds., 2012.

62 Which is not to say, "lacking in concreteness." Colonialism and its direct repression of protests to protect the land (e.g., The Dakota Access Pipeline protest) and ongoing internal policing of the ghettos populated by descendants of former slaves (e.g., as refused in the Ferguson uprising) is violent and concrete. But the sources of its rationalizations, the equivocations, the ambiguities behind which accountability disappears and no one is to blame, by which something as diffuse as "the system" emerges as a problem, where the systematic patterns go unmoored from words and mechanisms of responsibility – these *are* diffuse. An environment of rationalized domination shapes the underside of the environment of modernity (cf. Mignolo, 2011).

63 Where objectification is intrinsically a form of domination, since the person is dominated by the practical use of them, against the autonomy of their moral judgment.

64 I have removed the quotation marks mentioning the name, because the use is acceptable given self-determination and the prospective responsibility of resisting our situation through constructing processes anthroponomously. In other words, a counter-sensed "Anthropocene" passing through the decolonial work of the anthroponomy criterion is the Anthropocene as it makes autonomous sense to me.

65 See Chapter 2 of this novel.

66 See Gardiner, 2012 for analysis of the immoral and unjust dynamics, yet without awareness of internalizing colonial assumptions about rationality, including abstraction from the more than human world.

67 Coulthard, 2014.

68 Mignolo and Walsh, 2018.

69 Cf. Ngugi wa Thiong'o, *Decolonising the Mind: The Politics of Language in Afri-can Literature*, New York: Heinemann, 2011; but see also Shiri Pasternak, *Grounded Authority: The Algonquins of Barriere Lake Against the State*, Minneapolis: University of Minnesota Press, 2017 – part of the revisionist movement of critical legal schol-arship, including Andrée Boisselle and Sarah Deer, that shows the depths to which coloniality of languages enters systems of law and jurisprudence, thereby appearing as in-the-flesh and on-the-land power effects. See also Deer, 2015.

70 Tuck and Yang, 2012, p. 19: "Until stolen land is relinquished, critical consciousness does not translate into action that disrupts settler colonialism."

71 See especially Waziyatawin and Yellow Bird, eds., 2012. See also Taylor Sheridan, *Wind River*, Los Angeles: Thunder Road Pictures, 2017, a film that links fossil fuel extraction, geopolitical neo-colonization, settler colonialism, patriarchy; the slow vio-lence of drug abuse, limited economic prospects, and cultural loss and extinction, all to violence against indigenous women and the limited and under-resourced powers of tribal governance and self-protection under Federal Indian law. However, the film is colonizer-centered in its hero-savior, maintaining a film of coloniality.

72 In Chapter 4.

73 Notice the aspect switching going on in these sentences between the mention of the "Anthropocene," which is meant *colonially*, and the use of the word without quota-tions, which is thus meant *decolonially* to involve self-determination and, more gener-ally, anthroponomy.

74 See my *The Ecological Life: Discovering Citizenship and a Sense of Humanity*, Lan-ham, MD: Rowman & Littlefield, 2006, lecture 6; also my "Living Up to Our Human-ity: The Elevated Extinction Rate Event and What It Says About Us," *Ethics, Policy & Environment*, v. 17, n. 3, 2014, pp. 339–354; on wantonness, see my "The Sixth Mass Extinction Is Caused by Us," in Thompson and Bendik-Keymer, 2012, pp. 263–280 and "Species Extinction and the Vice of Thoughtlessness: The Importance of Spiritual Exercises for Learning Virtue," in Philip Cafaro and Ronald Sandler, eds., *Virtue Eth-ics and the Environment*, New York: Springer, 2010, pp. 61–83.

75 Gardiner, 2012; Mann and Wainwright, 2018; see also my 2016.

76 Whyte, 2018.

77 On the "social-ecological," see Dryzek and Pickering, 2019; cf. Vogel, 2015, on the socially constructed "environmental."

78 In 1994, Marie E. Edesess once told me, over lunch between work at the Legacy Coali-tion High School in Manhattan, NY and advocacy to stop the genocide in Bosnia, that the first act of social justice is to name "it" – the wrong done and the right needed.

79 Cf. Walter Benjamin, "Theses on the Philosophy of History," in *Illuminations: Essays and Reflections*, translated by Harry Zohn, New York: Mariner Books, 2019, pp. 196–209.

80 On the moral (second-personal consideration) and the ethical (third person consid-eration about flourishing), see my "The Moral and the Ethical: What Conscience Teaches Us About Morality," in V. Gluchmann, ed., *Morality: Reasoning on Different Approaches*, Amsterdam: Rodopi, 2013, pp. 11–23; Darwall, 2006.

81 The deep time of geology sees species come and go, taking much note only during periods of mass extinction or when – as is the case of aerobic bacteria – they change the entire system of the planet. If we have changed a geological interval, we are unlikely to outlive it to some imagined future interval. This is what it means that the Anthropocene involves broken words that include a broken world! I draw these remarks especially from a workshop on mass extinction, co-designed with Paul Pinet, Colgate University, April, 2012.

82 Mignolo, 2011; Mignolo and Walsh, 2018.

83 That would be autochthony.

84 "Allochthonous," *Oxford American Dictionary*, Apple Inc., 2005–2017.

85 Mignolo, 2011.

86 Such as found in the Algonquins of Barriere Lake according to Pasternak, 2017.

87 See Chapter 4 of this novel; and Mann and Wainwright, 2018.

88 Cf. Gardiner, 2012 and Matthias Fritsch, *Taking Turns with the Earth: Phenomenology, Deconstruction, and Intergenerational Justice*, Palo Alto: Stanford University Press, 2018.

89 See my 2006, lectures 2–6, and 10.

90 Because the broad view here is structured morally, losing control over the land would not imply the moral right to "ethnically cleanse" or even violently subject those who have illegitimately controlled the land. The morality and the politics of a just transition according to the allochthony condition – which has been mentioned only in a general and necessarily broad way in this chapter – requires, obviously, a lot more conceptual work. Cf. Colleen Murphy, *The Conceptual Foundations of Transitional Justice*, New York: Cambridge University Press, 2017.

91 Cf. Whyte, 2018 on "collective continuance."

92 Dryzek and Pickering, 2019.

93 See Chapter 4 of this novel.

94 See Vogel, 2015, for a critique of some eco-centric positions in environmental philosophy as involving a hidden "ventriloquism" of social processes. I agree with Vogel's social constructivist critique of the environment indebted to a practice-based reading of the insights of Hegel and Marx. There is no such thing as the "land" devoid of our social construction of it as part of our world. This is, originally, a Kantian point. See Chapter 4 for my detailed and fine-point disagreements with Vogel's work.

95 Vogel, 2015; cf. Whyte's, 2018, emphasis on collective continuance as being found within all manner of societies, including settler colonial ones. One of the foci of Whyte's work is ongoing criticism of romanticized views of indigenous life, just as Vogel also is critical of eco-Romanticism as meta-philosophical and meta-ethical positions.

On the braiding of practical and relational configurations of the land, see my 2006, lectures 2, 3, 5, and 10.

96 See Chapters 2 and 4 and also my forthcoming, *Between Us: After Martha C. Nussbaum's Politics of Wonder*, New York: Bloomsbury, 2022 est.

97 Cf. Martin Heidegger, 2010. One of my fine-point disagreements with Vogel is to take issue with his reading of Heidegger which, it seems to me, represses the role of *Angst* – ontological anxiety – in the "groundless" condition of worlds of sense. Vogel represses groundlessness when he denies the negation that *gives rise to* social construction in the first place. The totalization of social construction is a form of Heideggerian inauthenticity. We have to let the relationships in, including our personifying being lost (the "X"). See my *The Wind ~ An Unruly Living*, Brooklyn, NY: Punctum Books, 2018 concerning "the void."

98 See Vogel, 2015 on the formation of the environment by practices. See Young, 2010, chapter 2, for the way in which we receive the results of prior practices as a given situation into which we thrown. See Whyte, 2018 for sedimentation's complexities. The term "sedimentation," as I've suggested, is phenomenological. See Edmund Husserl, *The Phenomenology of Internal Time Consciousness*, translated by James S. Churchill, The Hague: Martinus Nijhoff, 1964. On being "thrown" into a world, see also Martin Heidegger, *Being and Time*, translated by Joan Stambaugh and Dennis Schmidt, Albany: SUNY Press, 2010.

99 See my 2006, lecture 4 especially.

100 See my 2006, lecture 6. The argument here is that the sense of common humanity appealed to in, e.g., *The Universal Declaration of Human Rights*, supports the view that a socialized person does not simply waste life, that the use of life requires a good enough reason, i.e., thoughtfulness predicated off of a prior respect for life. In neo-Aristotelian terms, this is a view of developed "second nature," and in Wittgensteinian terms, it is a view of the "grammar" of being human.

101 See Whyte, 2018; Mann and Wainwright, 2018; Mignolo and Walsh, 2018.

102 See my 2006, lecture 10 on "integrationism," as well as lecture 5 on relationships with lands.

103 The modality of necessity here – "have to," "must" – derives not from some putative obligation to survive, but rather from the moral demands that structure being a person, where not finding the conditions to flourish in a land will involve, by implication, undermining human dignity and where not being thoughtful with life will involve a failure of basic moral socialization. See Darwall, 2006 on interpersonal accountability; on the conditions of dignity, see Breena Holland, "The Environment as Meta-Capability: Why a Dignified Human Life Needs a Stable Climate System," in Allen Thompson and my *Ethical Adaptation to Climate Change: Human Virtues of the Future*, Cambridge, MA: MIT Press, 2012, chapter 7; and on thoughtfulness, again see my 2006, lectures 6 and 8.

104 See Dryzek and Pickering, 2019, chapter 3, p. 36 especially on reflexivity's elements; Burnor and Raley, 2018, p. 49 on authenticity in moral judgment and reasoning; and cf. Dryzek and Pickering, 2019, chapter 5 on "open" and "dynamic" sustainability.

105 Compare Whyte, 2018.

106 On the distinction between the moral and the ethical, see my 2013 and Darwall, 2006. The ethical focuses on the flourishing of the agent, not on interpersonal accountability.

107 In this and preceding paragraphs, flourishing, autonomy and capability interweave. According to textbook distinctions, however, flourishing is more substantive than autonomy, and capability is a condition of either (cf. Burnor and Raley, 2018 and Martha C. Nussbaum, *Women and Human Development: The Capabilities Approach*, New York: Cambridge University Press, 2000). I agree that capability is a precondition of both autonomy and flourishing. However, I disagree that autonomy is less substantive than capability. To say that someone flourishes is to say that their lives are good. Any understanding of the good that has authority will be one that is justified, i.e., that makes sense. To achieve that sense, one exercises one's autonomy. Flourishing is thus grounded in autonomy and is a result of it.

True, one may *say* that someone's life makes sense to them but that they are not living a good life. Yet precisely this point opens disagreement, that is, autonomy-in-relationship – for why would you say your life makes full sense to you if it isn't good? Flourishing is a category of autonomy that opens *the* world underneath any given world and asks us to look hard and to engage critically with what we think makes sense to us. Does it really? Is it really good?

108 See Chapter 3 of this novel.

109 This formulation is meant to echo Wilfred Seller's famous concept of the "space of reasons" as opposed to the "space of causes." See William DeVries, "Wilfred Sellers," *Stanford Encyclopedia of Philosophy*, 2016; Wilfred Sellers, *Empiricism and the Philosophy of Mind*, 2nd ed., Cambridge, MA: Harvard University Press, 1997. To be precise, the "space of personification" is a *category* of the space of reasons, which also includes theoretical and practical reasons. The space of personification is the category of relational reasons.

110 On the two directions of this analogical relationship, which I've previously called "analogical extension" – extending our humanity outwards – and "analogical implication" – finding the more than human inside our humanity, as a guide to it (e.g., "She is *rooted* in herself") – see my 2006, lecture 4.

111 On the adverbial as designating an important possibility in relationships, see my "Common Humanity and Human Rights," in John Rowan, ed., *Human Rights, Religion, and Democracy* (*Social Philosophy Today*, v. 21), Charlottesville: Philosophy Documentation Center, 2005, pp. 51–64.

112 See my 2013 and my "Do You Have a Conscience?" *International Journal of Ethical Leadership*, v. 1, 2012, pp. 52–80.

113 Cf. Darwall, 2006. Darwall focuses on accountability between autonomous beings. However, personification relations produce accountability as well. Autonomy is just

our form of being positively free as beings with lives of our own. See my and Amy Linch's "Freedom, Friendship, and Love: Mutuality with Other Species," Human Development and Capability Association annual meeting, University College, London, 2019.

114   See the reading of the text by Skye Arundhati Thomas in the previous chapter.

115   See David Schlosberg, *Defining Environmental Justice*, New York: Oxford University Press, 2007 and his "Justice, Ecological Integrity, and Climate Change," in Thompson and my 2012, chapter 8. In previous public discussion, for instance at a session of the Western Political Science Association, I was skeptical of Schlosberg's claims. I now think that they can be worked out along the lines I am drawing here and retract my overall criticism of Schlosberg's approach. See my "Can eco-systems be subjects of justice? No, but . . ..: Schlosberg, Nussbaum, and structural injustice," *American Political Science Association Annual Meeting*, Seattle, WA, 2011.

116   See Whyte, 2018, on the way "interdependence" becomes caught up in the space of personification as special relationships of responsibility.

117   Dryzek and Pickering, 2019.

118   On adapting our societies, values, and ourselves, see Thompson's and my concept of "humanist adaptation" – sometimes called "deep adaptation" today – in our 2012, introduction ("Adapting humanity"). On "thinking like a planet," see Hirsch and Norton, 2012.

119   See Burnor and Raley, 2018, pp. 49–50.

120   See Chapter 1 of this novel for discussion of claims about the end of civilization as we know it.

121   Immanuel Kant, *Critique of Pure Reason*, translated by Norman Kemp Smith, New York: St. Martin's Press, 1965; Susan Neiman, *The Unity of Reason: Rereading Kant*, New York: Oxford University Press, 1994.

122   See Tuck and Yang, 2012.

123   See Chapter 1 and also Dale Jamieson, *Reason in a Dark Time: Why the Struggle Against Climate Change Failed–And What It Means for Our Future*, New York: Oxford University Press, 2014; also Gardiner, 2012 and my 2016.

124   Hirsh and Norton, 2012.

125   There are moral conditions in which it is crucial to go back on one's word; for instance, if one's word suddenly led one to have promised complicity in an immoral act or pattern. But these conditions do not obtain in the context of colonization, where treatises were broken out of plain selfishness and in plain domination of the colonized.

126   Cf. Pasternak, 2017.

127   Compare Whyte, 2018, on vicious sedimentation and insidious loops; but also Alain de Botton's Rousseauian work, *Status Anxiety*, New York: Penguin Books, 2004 – although de Botton does not engage with a historically sensitive look into the colonial roots of the specific kind of anxiety that he most often documents.

128   It is the orientation to adaptive restoration I would advocate. See Thompson and my 2012, section 1 on adapting restoration. Rather than advocate for a specific ecological strategy, I would advocate for restoring landed autonomy, beginning with decolonization. In Chapter 4, I suggested that this would be a *meta-restoration* of the land.

6 What could others make of anthroponomy and how can I support them?

Monday, Shaker Heights, Ohio, a land to which my society has no right, springtime

One thing is still completely unclear about anthroponomy. It is how *humankind – anthrōpos*[1] – figures in it. This is the main and clear disanalogy with autonomy, which concerns the self – *autos*.[2] To figure out how to involve anthroponomy in the course and prospect of my life, I need to clarify how it implicitly invokes humankind. Given that I've been thinking of anthroponomy as a coordinated use of autonomy – even of nested, fractal forms of autonomy–I have some intuitions to pursue.

The concept of anthroponomy is supposed to help me consider autonomy-in-relationship as a decolonial process accountable with our planetary situation. But it is ambiguous how humankind ought to be involved in a manner consistent with autonomy-in-relationship. Is the suggestion that humankind is the *patient* of our processes insofar as they could have unintended, aggregate effects on the scales of the planet?[3] Or is the suggestion that humankind is a kind of *agent*, that is, that people all over the world and across time are to engage in parallel processes, a kind of fractal form?[4]

One thing that strikes me is that the former seems consistent with decoloniality, but the latter seems at odds with the spirit of self-determination in decolonization. In the former, moral accountability primarily involves delimiting social processes against wantonness. They must not spill out and affect people around the planet and across planetary time in any way that would undermine people's autonomy. Such delimitation also includes protecting the more than human world from wantonness, entrusted to, or respected within, social processes that mind it thoughtfully. What these thoughtful processes are and how to engage in the complex disagreement about how exactly one can be thoughtful with the more than human world is crucial,[5] but it is beside this general, primary point. To delimit wantonness from our social processes is to make sure our society's unintended effects do not undermine the autonomy of people around the planet and over planetary time, including the lands that support and are in relationship with thoughtful people. All of this is consistent with – and seems even implied by – a decolonial orientation that will challenge and undo the wanton effects of colonialism and its ongoing coloniality.

But the suggestion that anthroponomy implies a general form for social processes that must be internalized by people all over the planet seems complicit with, or even more an expression of, coloniality's global totalization of what it means to be human.[6] Some postcolonial theorists develop their concepts *opposing* just this. Gayatri Chakravorty Spivak, for instance, calls what resists such global totalization, "planetarity." Here, the planet serves as a figure of the abject and the erased, but also the resistant, unknown, and the inassimilable.[7] The question I want to address first is whether anthroponomy is "totalizing," and, relatedly, what is at stake in colonial totalization of the understanding of the human. What does that mean and imply? And is anthroponomy in the sense of indicating humankind as an agent of sorts, not just as a patient, a form of such totalization? Is anthroponomy in the agential sense[8] itself colonial?

The question is what it means to "totalize."[9] As I understand it, to totalize the human is to (a) occlude moral accountability[10] and to (b) practically or theoretically objectify people.[11] The notion of "totalization," including its expression, comes from deconstructive morality (incorrectly called "deconstructive *ethics*").[12] This morality is a form of *personalism*.[13] People deserve moral accountability. But "totalization," by turning people into objects, cloaks moral accountability.[14]

There is more to the idea, however, and it depends on a view of human being as *ontologically free* and *anxious*. To understand what this means, however, one has to understand anxiety, for instance, differently.[15] The idea, as I understand it, is this: we humans are weird. This means we are fated,[16] for instance by the world we inherit and in which we are raised. Yet our being fated also means that we are undetermined to a degree. The ways in which we come to terms with our world – make sense of it – involve us in interpretation and re-interpretation of it. And thus the world becomes open, slowly, along the path into which we are "thrown."[17]

The openness of the world, however, is not a complete liberty. How we make sense is from how we have learned to make sense. We work within a weird space – neither completely wide open nor completely fixed. We move through the drift of sense and non-sense, sometimes coming to terms with shifts we ought to make by coming to terms with what really does make sense to us in the face of the mystery of the world, including what appears through disagreement.[18] Sometimes we reject the sense we've been given as non-sense, as I did with "liberty."[19] Realizations are had, and I had them. We can deliberately construct the world otherwise to a certain degree.[20] The process is highly conflictual, contingent, and collective, but most often as an aggregate and uncoordinated collective whose processes start to cascade at various points into new discrete norms or aspects of ways of being, sometimes even into recognizably new ways of being.[21]

To live in this weird condition, fated and partially open, is to live in anxiety. This is not a comment about how people feel.[22] Anxiety as an ontological condition can appear in all sorts of ways in different cultural worlds. In a society of self-possession such as colonialism is, anxiety may appear as a horrifying void.[23] But it could also appear as wonder or as curiosity.[24] Some of these manifestations of anxiety – horror – shut down reflection, and some others – wonder and curiosity – open up reflection, albeit within ways that are delimited in advance, thus being fated, and yet can produce surprising shifts.[25] Forms of anxiety are often heteronomous, but some are autonomous. Nonetheless, in both kinds of cases, anxiety is the condition of sensing that the world may not make sense, which is to say that it could make more sense in some other way.[26] And why don't we essay that becoming?

Colonialism is especially vicious within such an ontological context, because it tries to fix the world through domination. It is a form of deeply heteronomous anxiety.[27] If colonialism were a monologue, it might say: "The world could be another way. So we will dominate and exploit it, totalizing the potentially free people in its midst and turning morally charged lands into mere resources, objects for the taking." Domination does violence by keeping people down and by bringing people up to be dominated and to reproduce domination. It completely twists people around if it keeps them alive at all.[28]

"Totalization" is the quality of dominating worlds that excludes the openness of people to re-interpret their world as they think makes sense. In this, totalizing systems keep people "in their place"[29] and delimit the capacity of people to reassess their world self-consciously and together. Free of domination, people are normative and uphold, quite often un-self-consciously, yet also self-consciously,[30] norms together. In a world without domination structuring its fate, sticking to a given world comes to be a partial result of autonomy-in-relationship. That world makes sense, and so we stick with it.

Finally, there is a third meaning of "totalization," one concerning not people but systems of sense.[31] According to this meaning, systems of sense "totalize" what we find sensical (understandable) and sensible (practical and relational), thereby excluding – and in this case repressing – the "others" of sense. These others are different ways of making sense and of being sensible. Totalization in this third meaning is tantamount to a failure of imagination built into systems of sense – discourses, practices, institutions that embody and govern what make sense, such as academic disciplines, methodologies and canons, and universities do. This failure is hidden in the normalcy of the system, its purported authority and appears in habits of sense-making where writers, scholars, speakers, etc. routinely exclude not only other ways of making sense in the world but the fact of everyone's limitations and historical particularity when making sense in the totalizing systems.

For obvious reasons, this third meaning to totalization is conducive to coloniality, so much so that it is often taken to be identical.[32] For instance, when colonialism erases the pre-colonial meaning-making of indigenous ways of life by relegating it to *pre*-civilization, it appears to totalize in this third sense.[33] The totalization of sense can erase cultures and set the way for the legitimation of colonialism. When it does so, it is a form of coloniality.

The question is whether anthroponomy engages in any of these three meanings of totalization. On reflection, I don't think that it does. First of all, anthroponomy is grounded in personalism, as autonomy-in-relationship is. This makes it opposed to the first meaning of totalization. It cannot totalize the human in either sense of denying moral accountability or of objectifying people.

Secondly, anthroponomy involves people everywhere realizing the anxiety of being human in an autonomous way without dominating people anywhere and without turning the morally charged world of life into a mere object for use and abuse. So the second meaning of totalization doesn't apply to it. Anthroponomy is *predicated* off of people's moral judgment and making sense of the world on their own terms.

Finally, as a form of autonomy-in-relationship, anthroponomy is so pluralistic as to lead to acknowledgment of the pluriverse, one grounded in the mystery of the world. This makes it break apart totalization in its third meaning. Any order of knowledge that claims to have no space for disagreement and nothing to learn from other systems of sense is always already heteronomous according to the autonomy-in-relationship of anthroponomy.[34]

Considering, then, the worry that anthroponomy is another form of colonial totalization, anthroponomy actually looks more like postcolonial "planetarity" resurging not just as a negation but also as a *position*.[35] In other words, anthroponomy appears *within* the decolonial space of the concern, on its side.

I think that this realization clears me to say something more specific about anthroponomy as a planetary and fractal form. When I can sit down again to write, I will do so.

Tuesday evening a week later, Shaker Heights, Ohio, a land to which my society has no right, springtime

In the past week, I've done some diagramming and outlining in order to arrange my thoughts. Anthroponomy is a *specific ordering* of autonomy-in-relationship, a coordinated use. That is my main idea about anthroponomy. The ordering is *teleological*, and it is *mereological*. Moreover, both its teleology and mereology have two dimensions – one spatial and the other temporal.

Mereology is the study of part-whole relations. A mereological consideration is one where a part is seen in view of a whole, a whole in view of its parts, or parts of a whole are seen in relation to each other.[36] To say that anthroponomy has a mereological ordering of autonomy-in-relationship is to say that anthroponomy considers autonomy-in-relationship in terms of parts to a whole or of parts of that whole in relation to each other. In this suggestion, what then are the parts, and what is the whole?

My idea is that anthroponomy is a coordinated use of a fractal form of autonomy-in-relationship where relationships of autonomy are parts of a wider whole wherein these relationships are consistent with the autonomy in each other relationship. This whole reaches as far as humankind and is structured by the consistency demands on the *outside* of any relationship-in-autonomy. The spatial dimension of this mereological ordering is planetary, and the temporal dimension is deep time.

Spatially, the *whole* of planetary, anthropogenic causality must be considered by the *parts* of anthroponomous action in any locale. When we aim for autonomy in our community, we must be mindful of the ways in which our social processes are stitched into what is happening all around the planet, affecting it or participating in it. We must coordinate what we do, with our effects around the whole planet as our horizon, including the effects of social processes that fundamentally change Earth.

Temporally, the *whole* of our effects into deep time must be considered in the *parts* of human history in which we participate. We must aim to coordinate what we do across time in light of the outside of the whole of human history, that is, with the limit of our own extinction and of a mass extinction as our horizon. Anthroponomy is mereological with reference to the planetary scales on which it must seek to coordinate people's autonomy.

Teleology, by contrast, comes in to explain the specific point of anthroponomous coordination. Teleology is unclearly defined in conventional English.[37] But as a term of art in conventional philosophy education,[38] to say that one is after a teleological consideration is to say that one is after *the ordering of something toward a purpose*. When the thing is an action, for instance, its teleological ordering is an intention.[39] When the thing is a living being, its teleological ordering is a living form.[40] Social processes have teleologies, too, or they would not be processes.[41] A process has a beginning, a middle, and an end. It has a form and a point. To say that anthroponomy involves a teleological ordering of autonomy-in-relationship is to say that anthroponomy involves autonomy-in-relationship within a purpose-driven process. In this suggestion, there is a purpose to autonomy-in-relationship that must be involved in it for it to be anthroponomic. What is the purpose?

The idea I'm considering is that anthroponomy is a coordinated use of autonomy-in-relationship with the purpose of *becoming collectively accountable for our planetary situation*, especially our social processes and their global and intergenerational effects far into the future. The goal here is to organize autonomous lives together. Anthroponomy is a specific use of autonomy in a social process of its own, one that seeks to coordinate people and their relations enough to protect the autonomy of humankind around the planet and far into the future, including humankind's relations with their lands. Therein are the spatial and temporal dimensions of the teleology, too. The mereology of anthroponomy is part of its purpose.

Given that the idea of anthroponomy is so abstract, it's worth asking why it should be abstract – what the use of that is. The answer is that the idea is regulative.[42] It is an idea, and it can never be definitively realized for reasons I'll discuss shortly. Because of this unrealizability, the idea serves as a critical challenge that produces an *un*settling process. Oriented by the regulative idea of anthroponomy, one cannot foreclose the moral demands of autonomy anywhere or anytime by presuming that one's world is settled. The demands of self-determination anywhere or anytime, consistent with moral accountability to others, pressurizes the scene of one's life, making it constitutively open to disagreement-in-relationship and the demand to be thoughtful with the more than human world.[43]

This unsettling, of course, is found in autonomy itself. What makes anthroponomy a specific use of autonomy is the coordination function of it. In the process of being constitutively open to the moral demands of the autonomy of others (something that is implied by autonomy anyway[44]), engaging in anthroponomy means orienting oneself to coordinate with others so that the lands which compose the planet's pluriverse of worlds sustain the landed autonomy of people and their more than human relations on planetary scales of space and time. Yes, much of this is implied by autonomy, but the function is a specific use of autonomy and a set of specific relations, a practical and relational focus to autonomy. The function serves as a coordination condition that brings the processes subjected to anthroponomy into focus.

In other words, having good relationships with people anywhere and every-where must pass *through* moral accountability, including accepting the perpetual possibility of disagreement (this is the general demand of autonomy). It must involve being committed *to* passing along (temporally) and sharing (spatially) the more than human world in a thoughtful way (this is the coordination condition of anthroponomy as a specific focus of autonomy).[45] In everday, plain words, good relationships depend on being open to people considerately and being thoughtful with the world you happen to share. It is our planetary situation given to us by the wantonness of colonialism, capitalism, and industrialism, that requires coor-dination. A specific coordination of good relationships involving moral awareness of the whole of humankind on Earth across space and time, anthroponomy is opposed to all social processes keeping people from being morally accountable to each other in sharing life on Earth by living within lands. Anthroponomy's rela-tionships are in this way unsettling. They are decolonial.

Accordingly, it's not hard to understand why anthroponomy cannot be finalized and is an open-ended process. To imagine it could be finalized would be to enclose it against others. But moral accountability ordered by attention to the planetary dimensions of our social processes perpetually opens us to the question of whether the world our social processes produce undermines the autonomy of others now on this planet or in the future.[46] In technical terms, such moral accountability opens us to question of whether we are directly dominating others[47] as well as to whether, indirectly, we participate in structural injustice affecting others now or in the future.[48]

At the same time, anthroponomy is historically rooted. The specific ordering of anthroponomy deals with the history of European colonialism, capitalism, and industrialism that has materialized the claim that "injustice anywhere [can be] a threat to justice everywhere."[49] In the last half millennium, a specific locale of humankind produced wantonness with the more than human world and under-mined the autonomy and self-determination of people around the planet into the far future. That injustice somewhere has become a threat to justice everywhere – and far into the future. Such is the story of European colonialism and British capitalism.[50] To orient ourselves by anthroponomy is to remain accountable to the historical actuality of such a planetary wrong as has befallen us and those who come after us, too.

Anthroponomy brings into focus an aspect of the human condition too – our fat-edness with respect to the past. We are thrown into existence inheriting a world[51] in which we are fated to become accountable if we are to become autonomous with others with whom we or our descendants will share the world in the future. Anthroponomy takes this human condition in freedom and scales it to the planet, providing a general social process driven by a critical, regulative idea. This social process aims to involve accountability for the demands of making sense in the world as a planetary, interconnected environment involving our basic moral accountability to each other. In the Anthropocene, understood as a prospective responsibility to become anthroponomous, to be human is to be fated to become accountable to each other across planetary space and time.

Thursday evening, Shaker Heights, Ohio, a land that should be unsettled in a house where I am unsettled

The almost cosmic abstraction of anthroponomy still bothers me. As it was at the beginning of this process almost a year ago, I still need to bend anthroponomy toward the plainly relatable. I think back to my first impulse, which was to engage in community politics. That located me where I live right here in a relatable way. What has become of that impulse?

Anthroponomy seems to involve a subtle reorientation to life on this planet, rather than a cosmic fix for all our woes. It's important that I keep in mind its subtlety in order to know how to involve it in my living. What I've realized over the past months could be a helpful guide. I learned that I should disagree in relationship with others,[52] confronting colonialism first and foremost by seeking to establish good relationships with all others.[53] In so doing, I should oppose in particular moral wantonness with lands and with others, including broken words.[54] Those whose relations to their environment violate the self-determination and autonomy of others, who are thus affected allochthonously, lose their right to the land. These realizations give me specific things to set in motion in my community.

What I can now add to them is that my moral attention ought to pay special attention to the ways in which my community plays a part in a larger whole. Does it contribute to, avoid, or oppose social processes that risk undermining the autonomy of others around the planet and far into the future? In other words, I ought to be aware of how my community plays a part in defeating or supporting the self-determination of others now around the world and into the far future, especially through its socially produced environment. And that will not be all, because I must be looking to coordinate my community with others.

Is my community defeatist of others, especially by not even having a land in which we are thoughtful with the more than human world over time? Then I ought to work to hold my community accountable, starting by refusing to acknowledge its right to the land in which we live. Is, on the other hand, my community supportive of others, especially by belonging to a land in which we are thoughtful with the more than human world over time, providing others in the future the conditions for an autonomous life? Then I ought to work through my community to support its part in a whole world where social processes become open and morally accountable to autonomy coordinated to account for our planetary situation.

My goal here should not be fixed. My task is unrealizable insofar as the process of anthroponomy is open-ended and open-bordered, so to speak. Still, I can see a general orientation to living. I want to make it illuminate more than that it already has.

Saturday morning, Shaker Heights, Ohio, a community without anthroponomy

I do not have a list of lifestyle changes I should make when I involve anthroponomy in my life, but I do have a clear thought about how to go about actually living with what I've been realizing. My thought is that, while there is some out and out personal viciousness in our planetary situation,[55] the main part of our situation has become structural.[56] For the great part of humankind, the problem that plagues us is one of structural injustice, not *our* direct viciousness regarding others around the planet or in far future generations. In the current wantonness of globally interconnected societies, it is the unintended consequences of our recurring, widespread practices that are a challenge to anthroponomy.[57] If I am to involve anthroponomy in the course and prospect of my life, I must consider how to become accountable for these practices as just one person in a community, including their relation to colonialism.[58] My understanding of anthroponomy's highly abstract mereological and teleological regulative idea must involve these more discrete, problematic conditions. This is how I can specify my purpose more.

Let us say that I work in my community for it to take up social processes that respect the self-determination of people with landed autonomy beginning with the colonized. Alongside any indigenous movement that so determines itself, I work on my end for actual decolonization. I refuse to continue broken words. Moreover, I refuse to recognize the right of any claim to the land if the claimant is wanton with the land. Within these things that I do, how should I understand my work in relation to the planetary task of anthroponomy?

The first thing that jumps out in such a question is that colonialism has been and still is global. Moreover, its extractive history – its relation to the land as a mere resource for the wealth of colonialists and the functioning of our fossil fuel economy – has made colonialism planetary. To seek decolonization is to seek to undo a world order that continues its domination in multiple parts of the globe[59] and which in its underlying dynamics and worldview is putting at risk people and lands on planetary scales of space and time. It is no small thing to challenge colonialism still today. To do so already does much to advance anthroponomy, because its functioning is global and planetary at once, involves domination, reproduces senselessness, and propagates wantonness with the planet's living order.

If we expand our focus to include coloniality in addition to ongoing colonization, then we take in capitalism, industrialism and the information economy as well.[60] These are economic forms that, despite their considerable variety, proceed with a wanton supposition, namely, that the land is an abstraction – some way to gain capital, some resource for mass production, or something irrelevant to virtuality.[61] In each, the places of relationship between people and with the more than human world are continually tested and eroded in the interest of, for instance, return on investment, efficiency, or flow.[62] The relation to the land – reducing it practically or erasing it (which amounts to the same thing, relationally) – is part and parcel of the objectification of people. To challenge coloniality in the major economic guises of my society – capitalism and industrialism especially – is to engage with global, systematic processes that have become dangerous on planetary scales, imperiling the autonomy of people and disrupting lands as thoughtful relations with the more than human world around the planet and deep into the future. In this sense, anthroponomy demands that I be anti-capitalist and anti-industrial as dimensions of being anti-colonial. In the place of capitalism and industrialism, I must work for economies morally accountable to people and thoughtful with lands on planetary scales. And the "information" that matters is that still found within the land.[63]

What would these decolonial economies be?[64] They need not be reactionary,[65] or even revolutionary, but they are in some sense evolutionary. To revolt is to roll back things. But to evolve is to open them out.[66] Opening out our economies makes sense. What is at stake is reordering political economy to become anthroponomic, especially though land relations that bring in the planetary reaches of space and time needed to mind the indirect, unintended, aggregate effects of colonial social processes around the globe, capitalism and industrialism in particular.[67] This reordering may, for all practical purposes, require – as a practical consequence of a moral demand – technological measures involving "ecological rationality"[68] that, for instance, separate the "technostream" from the "biostream"[69] to stop burdening the latter with the unintended effects of the former. It may require a new use of increased forms of energy,[70] not necessarily extractive,[71] but – for instance, in new energy generation and delivery models.[72] These, of course, are just general possibilities. But in all these measures, the economy needs to open out to the land in its planetary reaches.

The point to focus on is the moral pressure that the idea of anthroponomy places on judgments about what we ought to construct together through our social processes. Anthroponomy keeps moral accountability on planetary scales in view as a response to the planetary situation in which we are thrown in colonialism. What it generates as a response is an *orientation* to our political economies that refuses the land abstraction and dominating logic of coloniality. In their place, the anthroponomic orientation demands that we shape our societies in such a way that we can be morally accountable to the autonomy of others around the planet and into the planetary future, especially through our land relations. The issue here is not whether to be "technological" or "economical," but what these exactly mean. If they are anthroponomic, then they must be morally accountable in a way that coordinates people's autonomy enough to bring into account the effects and constitution of social processes on planetary scales so that they aren't wanton. That is the purpose of the pressure. The pressure here is to create, together through disagreement-in-relationship, anthroponomic technology and economy.[73]

In specific and precise ways, then, anthroponomy is anti-capitalist and anti-industrial. It does not court a nostalgic and idealized past, nor does it seek revolution that rolls society back. Instead, it seeks to open economy up in moral accountability. This could be thought of as the inclusion of humankind into economy, whereas currently existing capitalism – structured still by colonialism and predicated off of extractive industrialism imperiling future generations – is fundamentally exclusive, built on domination, and negligent to the point of moral crime.

As my argument over the past months has concluded, economies and their implicated technologies must be coordinated differently, open to the disruptive and patient social processes of disagreement-in-relationship. Part and parcel of this is that decolonization is necessary too. Decolonization involves delegitimizing land stewardship in any case where land relations do not obtain, thereby throwing into doubt the property rights of the still colonial world order and actually existing capitalism.

What would it take for land to no longer be abstract property but a field of morally accountable relations inclusive of humankind? Anthroponomic economies and technologies cannot be capitalist. The land relations undermine the abstraction of capitalism and its basis in property without land relations. They also cannot be industrial – for instance, merely practical with the land.[74] They must be something else, something novel and accountable at the same time with indigenous land relations.

A simple way to put this conclusion is that anthroponomy generates a moral demand for a new kind of economy and a new kind of technology after colonialism and coloniality. It may seem odd to speak of a new *kind* of technology or economy, but only if we do not see each as social processes formed by and into specific social forms. To not do so would be, once again as happened with the name "Anthropocene," to reify them,[75] making them unaccountable to social decisions and social construction as if they were physical forces and people were merely objects in their determined stream of causality. If technologies and economies are social processes, or part of them, and if social processes are structured *as social* by their moral or immoral dimensions, which they must by virtue of involving people,[76] then a change to the moral form of a process implies a new structure to the process. But a new structure to a social process implies a new kind of process, at least on the face of it.[77] Anthroponomic technology and anthroponomic economies are thus new kinds of things, at least from the standpoint of colonial technologies and economies.

It's in this sense that anthroponomy pushes toward something novel.[78] It helps seeing how so to put our ecological situation in perspective on a geological scale: We are in a time of extremely elevated extinction rates which risk the beginning of the sixth mass extinction since life began on Earth 1.5 billion years ago.[79] This looks to be a new geological situation, or so it very well could be. Moreover, it is a result of coloniality – of social processes that are thoroughly practical and given to domination of people and abstract use and wantonness with lands. The new geological situation, voiding life on a planetary scale, is a consequence of a bundle of globally interlinked social processes organized by coloniality as an immoral social form predicated off of wantonness with life and the domination of people who might stand up for land relations and thoughtfulness with life. Against the novelty of the planetary order of life and its "biological rules"[80] being thrown to the void, anthroponomy must demand of every economy, technology and polity whether they are ecologically reflexive with respect to the risk of mass extinction. If they are not, anthroponomy must be committed to make every economy, technology, and polity that affects the planet ecologically reflexive.[81] It must coordinate the autonomy of people to do so. To avoid further domination and wantonness, there mustn't be any acceptable social form in our planetary situation in any society that intentionally or unintentionally affects others on a planetary scale – or is part of the processes that so affect others – without that reflexivity. Anthroponomy demands thoughtfulness with life on a planetary scale for the sake of all people and for the sake of the myriad forms of life that exceed us in time, space, and understanding and that are also morally considerable.[82]

To involve anthroponomy in the course and prospect of my life demands being part of such work, novel to a colonial world. This conclusion fills out what "responsibility for justice" should involve in a global economy and still colonial world system structured by injustice.[83] Ours is a colonial world where most of us unintentionally contribute to processes that are incentivized and permitted to produce wanton results. The wantonness falls on people and the order of life around the planet and into the planetary future. It is predicated off of ongoing colonialism, bundled together with a set of social processes that coalesced five hundred or so years ago, at first, then two hundred years ago, in the Industrial Revolution, until the increasingly linked global order of capitalism and mass-scale industrialism "accelerated" the planetary order to new extremes of wantonness during the last three quarters of the twentieth century.[84] The "social connection" – the collective action in social movements – we ought to pursue anthroponomically – we the anthroponomists – must "connect" around dismantling the colonial world to reconstruct relationships that are anthroponomic and which undergird practices and then structures that are, too.[85] We must connect together with that larger whole in view.

Anthroponomy is a philosophy of good relationships, something that attention to colonialism has revealed as central to the moral issue at stake, and the relationships place new demands on us all who participate in social processes affecting the planet. The response to the domination and wantonness that structure the still colonial world must not be seen merely as a practical coordination, but as beginning in a new emphasis on politics through relationship.[86] It is the good relationships that come first, foremost through, in and across lands where thoughtfulness with life open to the planetary whole and aiming for planetary accountability are woven into relational practices of interpersonal and communal accountability.[87] From within the order of colonialism, capitalism, and industrialism that we inhabit, anthroponomy's prioritization of relationships is new in that it is planetary and demands decolonial work at the same time. It breaks through our extant global order. Drawing on indigenous land relations, it isn't new in spirit.[88] What is new is only the planetary ordering relations called for by our perilous situation now. In this way, anthroponomy makes sense as an evolving use of autonomy coordinated within the demands of our critical time.

I am getting closer to where I want to be so as to relate to my situation clearly. But before I finish my work of this past year, I should sit with where I've arrived for a while, then come back and write.

Two weeks later, late spring, early evening, Cleveland, Ohio in my university office once of the "Western Reserve," a land deserving now to be unsettled in light of planetary demands

Anthroponomy contextualizes relationships in specific ways, giving them ethical, moral, and political shape. In an anthroponomic orientation toward my life,[89] I work to live in a world that isn't senseless between people around the planet and across time far into the future. This work appears under the pressure of a regulative idea, since what it calls for is ideal and is far beyond the actual social processes in which I live. Still, trying to be accountable over the course of my life, I seek to construct social processes with others that will not stand to cause senselessness between people both here, now and around the planet into the far future.[90]

Senselessness between people is, indirectly or directly, dominating in that people find themselves unable to live a life in community that is open to their wonder, concern, or disagreement. When relations between people are dominating, the basic human demand to relate to the world with intelligence is closed, people's eyes "pushed downward" when they meet.[91] Refusing such a world, the pressure is on me to live thoughtfully, "thinking like a planet,"[92] about the effects of the social processes in which I participate on others now, here, and around the planet into the far future. These processes must create or conserve the conditions in which people can be thoughtful with each other.

At the center of these conditions is the land, understood as our environment involving our thoughtfulness with life as something that one does not simply waste or use without a good enough reason.[93] Wantonness with our environment destroys the conditions of sense between us. But the land in which we live is the place in which we can together make sense of our world. To lay waste to the land is then out of the question.[94] Use of the land by our social processes must be considerate of the lives of others around the planet into the far future and involving the past I have inherited, here in overtone time.

This consideration is not dogmatic. The social processes shaping our environment must remain open to disagreement in relationship – here in the present, anticipated or conjectured with the future, and studied with the past.[95] Involving anthroponomy in the course and prospect of my life, I must ask questions like, "How can my use of the land be part of a good relationship with others?" "How can I relate thoughtfully along with the land wherein I live?[96]" These questions must not be occasional eruptions of self-consciousness. They are to be part of my life. They demand a relational *attitude* of me, something that becomes part of my way of life and character.[97]

This attitude takes a question and makes it become a quest. I can live requesting with my being that things in this society change. That's called "having an attitude." Anthroponomy implies a decolonial attitude regarding the land and in the search for good relationships that are planetarily minded, aware of the plurality of worlds across space and time and refusing to produce wanton consequences for them as results of our world's social processes.[98] My questioning attitude is grounded in and focused on moral accountability.[99]

In turn, that moral accountability should lead me directly into *political* life where I seek to coordinate autonomy in practices, institutions, and regimes. Anthroponomy's teleological accountability implies that I am to live my life seeking through it to construct social processes that are accountable to autonomy-in-relationship through lands, and I am to do so by remaining questioning and critical of any social processes that appear to be wanton with our environment. I am to seek through our social structures good relationships involving lands around the planet and far into the future. Such a quest must be political, because political power orders our social structures.[100] The structural issues created through colonization and legitimated through the processes of coloniality – such as, capitalism, industrialism, patriarchy, racism, etc.[101] – are at the center of the issues to which my life and work must be accountable. They demand political change.[102] So my anthroponomic orientation must become political out of a questioning, moral attitude demanding thoughtfulness with our environment around the planet over time. That is what my moral commitment to protecting the possibility of sense between people demands.

What does my critical, political attitude lead me to push for? Surprisingly, in this world here and now, I cannot be free if I am not related politically to others in the intention to create anthroponomic processes.[103] If I am not so related, then my freedom is mere license,[104] not autonomy-in-relationship. There is no freedom in wantonness. Freedom as autonomy demands moral accountability, and in our planetary situation, such accountability implies anthroponomy. Autonomy-in-relationship in my society implies connecting myself up intentionally with planetary-scaled, political change, without which I cannot count myself free.[105] After all, I am involved in social processes that do affect others on planetary scales of space and time.

Moreover, given the weird world I have inherited, the main focus of my "daily request"[106] to make change must be decolonial, and that implies decolonizing.[107] Although abstraction is part of coloniality and its actually existing capitalism, humans do not exist in the abstract. We are historical, bound up in overtone time, and have inherited social processes structured by wantonness and injustice as a result of colonialism and its entangled social processes. The root of our planetary situation's problems lies in coloniality and the practices of economy and society constituting ongoing colonization alongside capitalism and industrialism especially. To be responsible in the Anthropocene must be to address this tangle with an anthroponomic orientation aiming for accountability at planetary dimensions. The thread of anthroponomy passes through the needle of decolonization and decoloniality. That has been what I have realized over this past year.

I should do two things at once in order to be politically engaged around decoloniality. First, I should cultivate my understanding of colonialism to understand more thoroughly how coloniality has a stranglehold on our world as it hurtles into a dangerous future in which the order of life we've inherited from the Holocene is at risk of mass extinction.[108] Part of this involves understanding what it means to support decolonization practically in my community – what treatises to demand be upheld, what laws to be challenged, what restitution is called for by the colonized.[109] Another part involves being able to trace out broad scale historical understanding about the construction of interrelated processes such as capitalism, industrialism, racism, patriarchy and even gender-binarism.[110] These are theoretical and practical aims, but they should show up in my capacity to orient my relationships toward moral accountability.[111] How can I be politically accountable if I do not understand colonialism?

Second, I should learn what is demanded of land relations on the temporal and spatial dimensions of the planet so that the social processes I help construct can be morally accountable. Whereas understanding colonialism emphasizes history, understanding our planetary situation emphasizes ecology and geology in addition. Whereas, politically, addressing colonialism emphasizes treatise honoring, reparations and restitution, addressing our planetary situation as it has been shaped by coloniality demands new forms of governance that are not yet part of humankind.[112] Decolonization and decoloniality go backward in time in order to go forward. Anthroponomy, involving decolonization and decoloniality, looks forward in order to become accountable for the injustice of the past and present while projecting a future open to the plurality of worlds in lands that aren't wanton. How can I be politically accountable if I do not understand our planetary situation?

To involve anthroponomy in the course and prospect of my life involves shaping a certain kind of world, accountable with other worlds in my community, around the planet, in the past, and – by confronting domination – into the future. Anthroponomy shapes a world in which my relating is *minded* in a certain sort of way – historically, ecologically, geologically and morally – and in which I seek to develop relationships in the contexts these frames of mind involve. In my relationships, I must (1) protect self-determination and question any social process that claims a land in a wanton way.[113] I must (2) work to construct or to protect land-relations that are morally accountable on planetary dimensions of space and time. Anthroponomy, involved in my life as an orientation, involves contextualizing my relations through moral accountability that is responsible (a) for my weird historical fate, (b) to people in their self-determination, and (c) regarding thoughtfulness with lands on planetary scales of space and time.

*

What I've been sketching involves something intimate to me. But I also think it should be intimate to any person who takes up anthroponomy. I have worked over nearly a year to bring the unwieldy, big picture idea of anthroponomy into view in such a way that I can own up to the Anthropocene in a sense that offers some moral clarity, a non-cynical meaning for the new name for our time, and a direction in which to work in daily life in my community. If I have that direction, I have something to do, and this is better than passive and delusional hope or paralyzed apathy. All of this work has been, in retrospect, anthroponomic, and it now takes me to something more personal than this fiction.[114] After all, it is time for me to conclude these desk and office meditations[115] and to step into something more relationally alive than this reflection.[116] The work of finding my way outside the word in the world has begun. The preceding reflections have helped, but their orientation is still wide open when I leave this office.

Anthroponomy involves an unsettling of every social process in light of a world we leave behind as a relation. Anthroponomy is a coordination of autonomy-in-relationship. Autonomy-in-relationship, in turn, is not a particular social process but a *way* in which social processes can become acceptable to people who confront each other in a world and are morally accountable. It is the general basis for sharing the world.[117] The core of this basis is being accountable to working through disagreement in relationship around the mystery of the world that appears in the forthright work of disagreement. Anthroponomy takes this core of a way in which people can confront each other and be together and coordinates it in light of the effects our worlds can have on others around the planet across vast reaches of time. It demands that we be thoughtful with the lands in which we can disagree and relate, in which the world can come to light between us. What anthroponomy primarily adds to disagreement-in-relationship is a prohibition on wantonness that is accountable with planetary dimensions of the effective power of our social processes. Such a prohibition is a deeper core to the way in which social processes can be acceptable to people now, here and around the planet far into the future. The idea is that the world people inherit and in which they find themselves comes with the thoughtfulness of those who are anthroponomic – not determining who they must be, but having left behind social processes that remain thoughtful with the world of life and open to people, non-dominating. What anthroponomic people thus leave behind is a world opened up. What is left behind for those who come after is a way of approaching the world that is open to the plurality of worlds and refuses wantonness and domination. It is the core of a relation – open to others, thoughtful with the world that is sharable. It's in this sense that anthroponomy "leaves behind a relation." It is also in this sense that it evolves.

As a way of approaching the world in its planetary and decolonial conditions, anthroponomy is unsettling. To remain open through moral accountability to others and thoughtful with lands on planetary scales eschewing and preventing wantonness is to not settle social processes but to leave them in a kind of anxiety. They must be open to challenge or they will be dominating. They must be open to ecological feedback and contest over thoughtfulness with life or they will be, latently or actually, wanton. They cannot settle or they will risk being arbitrary and thus mere license. Rather, any social process that I have developed over time with others and take to be anthroponomic must remain fluid, open to contest, feedback, and change. This fluidity involves "ecological reflexivity"[118] as well as a continually "adaptive" spirit,[119] but the source of this fluidity is not simply pragmatic.[120] It derives from the demands of autonomy-in-relationship. It is relational. When we are committed to relating morally to each other, including across lands in legacies we leave others far in the future and around the planet, our world cannot be settled and nor can we – nor can I.

The anxiety of anthroponomy, then, is different from the anxiety of the reified "Anthropocene" with which I began my reflections months ago. My anxiety right now after this year's anthroponomic work is not dread at the senselessness of things – the impossibility of possibility.[121] Rather, it is restless excitement over the possibility of relating and the need to keep being accountable. It is the space of moral duty. This anxiety is of a form of this world that must be ready to pass away and of a newly related world to come in light of disagreement.[122] Or, less grandiosely, the anxiety is implicit in being consistently considerate with others and mindful of lands, consistently not stuck in one's ways.

The excitement of anthroponomy's anxiety need not take a particular, emotional form.[123] Its way of approaching the world is intrinsically hopeful in that it anticipates disagreement-in-relationship and the social process of reevaluating social processes. But such a way of approaching the world can manifest in many emotional and temperamental ways. One might be humble, comedic, enthusiastic, or resigned while ironic. What matters is that, involving anthroponomy in the course and prospect of one's life, one expects to change the social processes of one's world in light of disagreement and improved thoughtfulness with lands. Moreover, one is ready in the anxiety, because morality calls for it.

I think of anthroponomy's *mood* as a constitutive excitement that is also a constitutive anxiety.[124] One cannot be anthroponomic without the unsettling nature of its orientation and the expectation of morally accountable engagement in changing one's world through good relationships. To be anthroponomic is to hover in the relating – which is also to anticipate the relating. *That* is what moral accountability involves.

This is the intimate thought I want to mark. It is also a thought that should, in some other unique manner, become intimate to any other person who considers anthroponomy. The anxiety of anthroponomy is a form of work toward moral authenticity.[125] In moral authenticity, we engage in reevaluating our values as a matter of moral accountability.[126] Work in anthroponomy is a species of work for moral authenticity in which we reconsider our social processes through our relations, over and over, never settled, across planetary reaches of space and time and in light of the plurality of worlds and the difficulty of being thoughtful with lands. The cores of this authenticity are disagreement in relationship and thoughtfulness with lands. Together, both create an ever-present possibility of counter-pressure to the acceptability of our social processes and thus lead one, if one is to be morally accountable, to engage with their reconsideration in light of anthroponomic demands. Here, being true to ourselves is being true to the disagreement, and being true to our lands involves the planet's space and time.[127]

Anthroponomy is an orientation toward, and a basis for, a specific kind of work in moral authenticity bound up with re-evaluating social processes in light of our planetary situation and the fate of inheriting colonialism on this planet. Its manifestations can take many forms, provided that the rough shape of the lives of anthroponomists and of anthroponomic processes involve the very general ordering relations I have been articulating in the past weeks. How I involve anthroponomy in the course and prospect of my life will be different than how you involve it, should you do so as you must. How we involve anthroponomy will be different than how others do, especially emerging from colonization around the planet or in the far future. To say that there will be difference is just to mark that autonomy-in-relationship proceeds from the differences between us, not the presumed sameness.[128] It is to mark that anthroponomy proceeds from disagreement-in-relationship and thoughtfulness with lands, both of which generate a continual pressure for moral authenticity worked out between us in light of worlds to come.

What I have engaged in this past year is a personal, anthroponomic process where the disagreement I've encountered is from the overwhelming presence of colonization still in my community and its linked coloniality shaping the discourse of planetary cataclysm that hangs over everyone's head today and for the foreseeable future. It's been *my* process, setting me up for an unsettling and more thoughtfully engaged future, one that circulates the past and allows its ghosts to be with my present as a power to change my world. But it has also been *only* my process. There must be others. This is a basic moral fact of differentiation.[129]

Tonight, after a year of searching, I have come full circle with anxiety. Yet I have done so in a spiral. I find that anxiety now is positive when one works with it, opening out into the world inclusive of humankind and our moral relations. Anxiety leads me to be inclusive in my thinking as I step from behind this desk and go outside. What could others make of anthroponomy? How can I support them?

These two questions form the line between the inside and the outside of my personal reflection this past year. They are my unsettling, and in them I will work for my society to change.

Notes

1 "Anthropo-" *Oxford American Dictionary*, Apple Inc., 2005–2017.
2 "Autonomy," *Oxford American Dictionary*, Apple Inc., 2005–2007.
3 Cf. Dale Jamieson, *Reason in a Dark Time: Why the Struggle Against Climate Change Failed–And What It Means for Our Future*, New York: Oxford University Press, 2014; Iris Marion Young, *Responsibility for Justice*, New York: Oxford University Press, 2010, chapter 2.
4 This was my initial suggestion in Chapter 1. See also my "'Goodness Itself Must Change' – Anthroponomy in an Age of Socially-Caused, Planetary, Environmental Change," *Ethics & Bioethics in Central Europe*, v. 6, n. 3–4, 2016, pp. 187–202; but also my "The Strange Un-agent of Our Species, Our Collective Drift," ISEE/IAEP Conference, Nijmegen, Netherlands, 2011.
5 Cf. my *The Ecological Life: Discovering Citizenship and a Sense of Humanity*, Lanham, MD: Rowman & Littlefield, 2006; *Solar Calendar, and Other Ways of Marking Time*, Brooklyn, NY: Punctum Books, 2017a; and *The Wind – An Unruly Living*, Brooklyn, NY: Punctum Books, 2018.
6 See Walter D. Mignolo and Catherine E. Walsh, *On Decoloniality: Concepts, Analytics, Praxis*, Durham, NC: Duke University Press, 2018.
7 Gayatri Chakravorti Spivak, "'Planetarity' Box 4 (*Welt*),*" *Paragraph*, v. 38, n. 2, 2015, pp. 290–292 and *Death of a Discipline*, New York: Columbia University Press, 2005, chapter 3. See Joel Wainwright's discussion of the *idea* (not concept) in his *Geopiracy: Oaxaca, Militant Empiricism, and Geographical Thought*, New York: Palgrave Macmillan, 2013, chapter 5.
8 To be precise, since anthroponomy is primarily relational, not practical, the sense is not precisely *agential*. The agent is part of practical reason (see G.E.M. Anscombe, *Intention*, Cambridge, MA: Harvard University Press, 2000). Given that anthroponomy is

primarily *relational* and that the *person* (or the *interpersonal,* and, collectively, the *people*) is the kind for *who* relates, anthroponomy in the sense I am indicating is the *personal*, not the agential, sense. But this will be confusing in ordinary speech. The distinction between who acts and whom is affected – between the agent and the patient – is a convention, albeit one developed from a tradition of "ethics" that is primarily practical and has not yet internalized relational reason as the core logic of the moral.

9 The *locus classicus* of this claim is Emmanuel Levinas, *Totality and Infinity, an Essay on Exteriority*, translated by Alphonso Lingis, Pittsburgh: Dusquesne University Press, 1969. But, with reference especially to postcolonial thought, a more influential intimation of the notion appears throughout Jacques Derrida's work, e.g., *The Margins of Philosophy*, translated by Alan Bass, Chicago: University of Chicago Press, 1982 or his.

10 Levinas, 1969, calls this the "face to face."

11 Levinas makes much of how theoretical reason avoids moral accountability. His focus is on phenomenological interpretations of theoretical reason. Practical objectification, however, becomes the issue once action is concerned. Cf. my "How Do You Approach Public Philosophy?" *Blog of the APA*, April 23, 2019.

12 See Simon Critchley, *The Ethics of Deconstruction: Derrida and Levinas*, Edinburgh: Edinburgh University Press, 2014. As I've marked in this essay, the logic of the personal is properly the *moral*, not the ethical (which concerns the *desirable*, and hence practical reason). See my "The Moral and the Ethical: What Conscience Teaches Us About Morality," in V. Gluchmann, ed., *Morality: Reasoning on Different Approaches*, Amsterdam: Rodopi, 2013, pp. 11–23 and, for support, Stephen Darwall, *The Second Person Standpoint: Morality, Respect, and Accountability*, Cambridge, MA: Harvard University Press, 2006. Critchley should agree in principle (see note 13).

13 Simon Critchley, for instance, understands that Levinas's "ethics" is a form of *personalism*. See Cliff Sosis and Simon Critchley, "What Is It Like to Be a Philosopher? Simon Critchley," *Blog of the APA*, April 16, 2019.

14 On "cloaking" what is morally demanding, see my 2006, lecture 8, "A Circle of Life."

15 The tradition here involves figures with profound differences, but primarily Sartre, Heidegger, Kierkegaard, and Schelling. Lacan is also important, although he is locked within practical reasoning's desire as a constraining logic. Rousseau, oddly, makes an appearance in his *Second Discourse*. The root of the tradition is a strong misreading of the neo-Platonism of Giovanni Pico della Mirandola and his often reinterpreted central passage from the *Oratio* in which he pictures "Adam" as ontologically unfixed and thus in need of using his own judgment to determine himself. Good thinkers within this tradition are (1) Jean-Luc Nancy, *The Experience of Freedom*, translated by Bridget McDonald, Palo Alto: Stanford University Press, 1994; (2) John Haugeland, *Dasein Disclosed: John Haugeland's Heidegger*, edited by Joseph Rouse, Cambridge, MA: Harvard University Press, 2013; and (3) Charles Larmore, *The Practices of the Self*, translated by Sharon Bowman, Chicago: University of Chicago Press, 2010.

Haugeland's reading is non-individualistic, however, and so strays far from many Sartrean interpretations of Pico's so-attributed idea. The most resolutely non-individualistic thinker, however, is Reiner Schürmann, who sees changes in "epochs of being" as out of the orbit of any agency. See Reiner Schürmann, *Heidegger on Being and Acting: From Principles to Anarchy*, translated my Marie Gros, Indianapolis: Indiana University Press, 1987.

Yet in all of these thinkers, Schürmann included, to be who we are is an essentially *normative* challenge whose ground is an ontological unfixity that is inseparable from having norms at all. The appearance of this unfixity in the space of meaning is called "anxiety." In recent work, I relate it to the logic of aesthetic judgment of the beautiful in Kant, that is, to the striving to find sense in the world in wonder. See Giovanni Pico della Mirandola, *Oration on the Dignity of Man: A New Translation and Commentary*, edited and translated by Francesco Borghesi et al., New York: Cambridge University Press, 2012; Jean-Jacques Rousseau, *Discourse on the Origin of Inequality*, translated by Franklin Philip, New York: Oxford University Press, 2009; F.W.J. Schelling, *Philosophical Investigations into the Essence of Human Freedom*, translated by Jeff Love and Johannes Schmidt, Albany, NY: SUNY Press, 2007; Søren Kierkegaard, *The Concept of Anxiety: A Simple, Psychologically Orienting Deliberation on the Dogmatic Issue of Hereditary Sin (Kierkegaard's Writings, v. 8)*, translated by Reider Thomte and Albert Anderson, Princeton: Princeton University Press, 1981; Martin Heidegger, *Being and Time*, translated by Joan Stambaugh and Dennis J. Schmidt, Albany: SUNY Press, 2010; Jean-Paul Sartre, *Being and Nothingness*, translated by Sarah Richmond, New York: Atria Books, 2019; and then my *Between Us: After Martha C. Nussbaum's Politics of Wonder*, New York: Bloomsbury, 2022 est.; my "Fleabag, Let Things Get Lost!" *Public Seminar*, September 19, 2019a; and "Lostness," University of Exeter, UK, September 12, 2019b.

16 See the root of "weird," *wyrd*, in Old English, meaning *fated*. "Weird," *Oxford American Dictionary*, Apple Inc., 2005–2017.

17 Cf. Heidegger, 2010 on "thrownness," his word for being weird.

18 On the drift of sense as it turns in a moment of personal realization, see my 2017, study 4, "I want to meet you as a person." Schürmann's (1987) notion of "epochal stamps" makes drift entirely impersonal, however. There are deep ontological disagreements about how the openness of our being (or our way of being, *ēthos*) appears.

19 See Chapters 2 and 3 of this novel.

20 Cf. Steven Vogel, *Thinking Like a Mall: Environmental Philosophy After the End of Nature*, Cambridge, MA: MIT Press, 2015.

21 One of my preferred sources for observing the drift of sense over long stretches of history was Philippe Ariès and George Duby, eds., *The History of Private Life*, v. I–V, translated by Arthur Goldhammer, Cambridge, MA: The Belknap Press, 1987–1991. Of course, this collection is positively colonialist in its European delimitation. And *that* realization, whereby today that historical work would not be possible in research academy due to its Eurocentrism, is itself a mark of drift, specifically, of decoloniality.

22 Cf. Heidegger, 2010.

23 Cf. my 2018, esp. stretch 1, "Vulnerability."

24 The subject of my forthcoming study– *Between Us: After Martha C. Nussbaum's Politics of Wonder* – takes up wonder in the *midst* of a society of self-possession to see what can be made of it in protest. On curiosity, consider Danielle Bassett and Perry Zurn, "Curious Minds: From Building Knowledge Networks to Inciting Political Resistance," MIT Media Lab, May 15, 2018 – including their *Curious Minds*, Cambridge, MA: MIT Press, forthcoming, and Zurn's *The Politics of Curiosity*, Minneapolis: University of Minnesota Press, forthcoming.

25 Hear especially Zurn in Basett and Zurn, 2018.

26 See my 2019a, 2019b.

27 Heidegger, 2010, would call it "inauthentic."

28 On the colonial fate today and being twisted around, see Ta-Nehisi Coates, *Between the World and Me*, New York: Spiegel & Grau, 2015. See also Jaskiran Dhillon, *Prairie Rising: Indigenous Youth, Decolonization, and the Politics of Intervention*, Toronto: University of Toronto Press, 2017. Think what it means to your world if you are the recipient of policies "killing the Indian to save the man." See "Institutional History," *The Carlisle Indian School Project*, accessed February 3, 2019.

29 On being kept in one's place by the "police" (Rancière's name for political totalization), see Jacques Rancière, *Disagreement: Politics and Philosophy*, translated by Julie Rose, Minneapolis: University of Minnesota Press, 2004.

30 I.e. deliberately, as endorsed on reflection. See Larmore, 2010.

31 It is this meaning that animates Spivak's (2015, 2005) use of "planetarity." It is similar to Derrida's scintillating and reappearing notion of totalization as the repression of finitude in our systems of sense. See Derrida, 1982.
32 Moments in Mignolo and Walsh, 2018 read this way.
33 Cf. Kyle Powys Whyte, "Settler Colonialism, Ecology, & Environmental Injustice," *Environment & Society,* v. 9, 2018, pp. 129–144.
34 See Chapter 2 for the working out of this point.
35 Spivak, 2015, 2005; Mignolo and Walsh, 2018.
36 Achille Varzi, "Mereology," *Stanford Encyclopedia of Philosophy*, 2016, accessed September 28, 2019.
37 See "teleology," *Oxford American Dictionary*, Apple Inc., 2005–2017, where the word is defined as "the explanation of phenomena by the purpose they serve rather than by postulated causes," and note that as of April 25, 2019, the *Stanford Encyclopedia of Philosophy* has no general entry for teleology.
38 Here, I draw on common sense as found within my scholarly formation and in existing scholarly communities including philosophy and the wider humanities.
39 Anscombe, 2000.
40 Michael Thompson, *Life and Action: Elementary Structures of Practice and Practical Life*, Cambridge, MA: Harvard University Press, 2008.
41 A process implies some kind of development toward an end. See also my "Development–Concepts and Considerations," in Willis Jenkins, ed., *The Spirit of Sustainability, v. 1 of The Berkshire Encyclopedia of Sustainability*, Great Barrington, MA: Berkshire Publishing, 2009, pp. 100–105.
42 See Immanuel Kant, *Critique of Pure Reason*, translated by Norman Kemp Smith, New York: St. Martin's Press, 1965; Susan Neiman, *The Unity of Reason: Rereading Kant*, New York: Oxford University Press, 1994. The sense of being guided by a regulative demand of reason also imbues the chapter on Odyssean happiness as striving, not settling, in her *Moral Clarity: A Guide for Grown-Up Idealists*, New York: HarperCollins, 2008.
43 On the role of thoughtfulness in anthroponomy, see Chapter 5 of this novel, as well as my 2006, *passim*.
44 See Chapter 2.
45 For the introduction of good relationships as a basic category of anthroponomy, see Chapter 3 of this novel.
46 It is not obvious how we can undermine autonomy in the past, but perhaps *failing to recognize the autonomy* of those in the past through various forms of misrepresentation would serve to point anthroponomy toward the past as well.
47 Cf. Eve Tuck and K. Wayne Yang, "Decolonization Is Not a Metaphor," *Decolonization: Indigeneity, Education & Society*, v. I, n. 1, 2012, pp. 1–40; Stephen M. Gardiner, *A Perfect Moral Storm: The Ethical Tragedy of Climate Change*, New York: Oxford University Press, 2011.
48 Cf. Iris Marion Young, *Responsibility for Justice*, New York: Oxford University Press, 2010.
49 Martin Luther King, Jr., "Letter from a Birmingham Jail," April 16, 1963. The text in question is: "Injustice anywhere is a threat to justice everywhere. We are caught in an inescapable network of mutuality, tied in a single garment of destiny. Whatever affects one directly, affects all indirectly."
50 Cf. Ellen Meiksins Wood, *Liberty & Property: A Social History of Western Political Thought from Renaissance to Enlightenment*, New York: Verso, 2012, chapter 1.
51 Heidegger, 2010.
52 See Chapter 2.
53 See Chapter 3.
54 See Chapter 4.

55 I think here of my current President's relation to the climate crisis, for instance. As an arbitrarian, Donald Trump is plainly vicious. See my "President Donald Trump Is an Arbitrarian, Not a Fascist," *Cleveland Plain Dealer*, August 18, 2017, accessed September 29, 2019.

56 Young, 2010.

57 See Jamieson, 2014; my 2016. Young, 2010, arguably touches on it.

58 Neither the Jamieson or my 2016 article considers colonialism.

59 E.g. The United States of America, Canada, Australia, New Zealand, Israel, to name a few locations.

60 See Moishe Postone, *Time, Labor, and Social Domination: A Reinterpretation of Marx's Critical Theory*, New York: Cambridge University Press, 1998; Geoff Mann and Joel Wainwright, *Climate Leviathan: A Political Theory for Our Planetary Future*, Brooklyn, NY: Verso, 2018; Manuel Castells, *The Rise of the Network Society*, 2nd ed., Cambridge, MA: Blackwell, 2000.

61 Recall Postone's (1998) discussion of state communist production as structured by industrial practices with people and to lands.

62 See Wendy Brown, *Undoing the Demos: Neoliberalism's Stealth Revolution*, Cambridge, MA: Zone Books, 2015 on the former and compare Herbert Marcuse, *One Dimensional Man: Studies in the Ideology of Advanced Industrial Society*, 2nd ed., Boston: Beacon Press, 1991 regarding the latter. See also Castells, 2000.

63 Cf. Shiri Pasternak, *Grounded Authority: The Algonquins of Barriere Lake Against the State*, Minneapolis: University of Minnesota Press, 2017.

64 See the question about "Climate X" as the unknown post-capitalist order in Mann and Wainwright, 2018, an allusion to Kōjin Karatani's famous fourfold analysis of modes of exchange, with the last – the "Kantian 'X'" – being anarchist. See his *The Structure of World History: From Modes of Production to Modes of Exchange*, translated by Michael K. Bourdaghs, Durham, NC: Duke University Press, 2014.

65 On reactionary responses, consider Jennifer Grayson's podcast, *Uncivilize*, 2018–19 or the "Preppers" movement, cf. Leo Benedictus, "Is It Time to Join the 'Preppers'? How to Survive the Climate-Change Apocalypse," *The Guardian*, February 17, 2014. There are many examples of responding to capitalism or industrialism by going "off the grid," "back to nature," or the like. The most infamous may be Christopher McCandless's in Jon Kraukauer's *Into the Wild*, New York: Anchor Books, 1997.

66 See "revolution," and "evolution," *Oxford American Dictionary*, Apple Inc., 2005–2017.

67 See my 2016 on the aim to reorder "forms of power" in our political economies to bring our unintended effects into moral accountability.

68 Cf. Steven Vanderheiden's "Informational Approaches to Climate Justice," in Aaron Maltais and Catriona McKinnon, eds., *The Ethics of Global Climate Governance*, Lanham, MD: Rowman & Littlefield, 2015, chapter 11 which draws on John Dryzek's use of the term.

69 On these two streams, see Michael Braungart and William McDonough, *Cradle to Cradle: Remaking the Way We Make Things*, New York: North Point Press, 2002.

70 Cf. Simon L. Lewis and Mark A. Maslin, *The Human Planet: How We Created the Anthropocene*, New Haven: Yale University Press, 2018. The two authors of this book have led the way on the scientific argument for the onset of colonialism marking the onset of our current planetary situation – which they call the "Anthropocene" – due to the continental pandemic and invasive species effects caused by colonialism. Nonetheless, they persist as most natural scientists seem to do and ignore social scientific precision with the "Anthropocene." Lewis and Maslin attribute causality to all of humankind rather than to the specific social processes of colonialism. Coloniality runs deep.

71 As is found in the pro-nuclear power movement. See James Lovelock, *The Revenge of Gaia: Earth's Climate Crisis & the Fate of Humanity*, New York: Basic Books, 2007.

72 I think here of Robert F. Kennedy Jr.'s advocacy for solar generation plants that generate the amount of energy of a nuclear power plant. Cf. his Town Hall speech, Cleveland, Ohio, 2013.
73 Consider the economies studied in David Schlosberg and Luke Craven, *Sustainable
Materialism: Environmental Movements and the Politics of Everyday Life*, New York:
Oxford University Press, 2019.
74 Here we can see somewhat clearly what separates anthroponomy from "eco-
modernism." True, eco-modernism does focus on the "decoupling" of the technostream from the biostream by way of "increased energy." But the anthroponomic base
of non-technocratic politics that autonomy-in-relationship implies, including its "ecological reflexivity" (John S. Dryzek and Jonathan Pickering, *The Politics of the Anthropocene*, New York: Oxford University Press, 2019), emphasizes decolonization, the
critique of capitalist property relations, and critique of industrial extractive industry as
a *condition* on what would count as a moral advance. Some aspects of these might be
interpreted into eco-modernism, but the most critical of these anthroponomic demands
are antithetical to the letter and spirit of the eco-modernist employment of capitalism
and advanced industrialism.
    That said, it is morally important to acknowledge what eco-modernism has under-
lined: that per capita resource use has improved in many ways through the growth of
modern technology. But it is also morally needed to assert what is left unstated: that
population growth and the rapid acceleration of the planet toward an entirely dangerous future for humankind and a next mass extinction are the results of the non-
anthroponomic, colonial social, political, and economic organization of the globalized
world. *These* must be critiqued squarely and fully, beginning with decolonization
and decoloniality. But eco-modernism conspicuously avoids the point. See J. Asafu-
Adjaye et al., *An Ecomodernist Manifesto*, April 2015, accessed September 29, 2019.
75 See Vogel, 2015 and Axel Honneth, *Reification: A New Look at an Old Idea*, with
Judith Butler, Raymond Guess, and Jonathan Lear, edited by Martin Jay, New York:
Oxford University Press, 2008.
76 Darwall, 2006.
77 Possibly, a new structure might indicate a species of a genus of a kind. But in ordinary
language, a new species of something is a new kind of thing.
78 See the preface to this novel.
79 The most current extensive – and highly disturbing – assessment is the IPBES (Inter-
governmental Science-Policy Platform on Biodiversity and Ecosystem Services) report
to the United Nations, May 6, 2019, which estimates 1,000,000 species currently on
the edge of extinction (accessed September 29, 2019). The report goes on to estimate
losses to *human* economies and capabilities. On the epistemic problems of determining
whether we are currently in the early phases of a mass extinction cascade, see my and
Chris Haufe's "Anthropogenic Mass Extinction: The Science, the Ethics, the Civics,"
in Stephen Gardiner and Allen Thompson, eds., *The Oxford Handbook of Environmental Ethics*, New York: Oxford University Press, 2017, chapter 36; see also Elizabeth
Kolbert, *The Sixth Mass Extinction: An Unnatural History*, New York: Henry Holt &
Co., 2014; my "Living Up to Our Humanity: The Elevated Extinction Rate Event and
What It Says About Us," *Ethics, Policy & Environment*, v. 17, n. 3, 2014, pp. 339–354;
and for a more personal relationship to the event, my 2017a. For the first extensive
study of the risk of mass extinction in our time, see Martin Gorke, *The Death of Our
Planet's Species: A Challenge to Ecology and Ethics*, translated by Patricia Nevers,
Washington, DC: Island Press, 2003.
80 See my and Haufe's 2017. Mass extinctions reorder the extant rules of life on a planet
by decimating the biological integration of the planet in the wake of the extinction of
massive amounts of its species.
81 See my 2014; Dryzek and Pickering, 2019.

82 See Paul D. Hirsch and Bryan G. Norton, "Thinking Like a Planet," and My "The Sixth Mass Extinction Is Caused by Us," in Allen Thompson and Jeremy Bendik-Keymer, eds., *Ethical Adaptation to Climate Change: Human Virtues of the Future*, Cambridge, MA: MIT Press, 2012, chapters 16 and 13. See also my "Species Extinction and the Vice of Thoughtlessness: The Importance of Spiritual Exercises for Learning Virtue," in Philip Cafaro and Ronald Sandler, eds., *Virtue Ethics and the Environment*, New York: Springer, 2010, pp. 61–83. Cf. Daniel Wildcat, *Red Alert: Saving the Planet with Indigenous Knowledge*, Golden, CO: Fulcrum Publishing, 2009.

83 Young, 2010.

84 See Lewis and Maslin, 2018.

85 See my 2016 on reconstructing anthroponomic "forms of power."

86 Cf. my "Democracy as Relationship," *e-Flux Conversations*, April 30, 2017b and my forthcoming monograph around the topic, est. 2022. Young, I believe, would support this emphasis on relationships, but it is, however, not emphasized in her 2010 posthumous work.

87 See Schlosberg and Craven, 2019; Whyte, 2018.

88 Whyte, 2018; Pasternak, 2017.

89 See my 2016 on the idea of an "anthroponomic orientation." On the idea of an orientation, see also my *Conscience and Humanity*, Ann Arbor: UMI, 2002, chapter 5 and Anthony Laden, *Reasonably Radical: Deliberative Liberalism and the Politics of Identity*, Ithaca, NY: Cornell University Press, 2001; and Sara Ahmed, *Queer Phenomenology: Orientations, Objects, Others*, Durham, NC: Duke University Press, 2006.

90 Of course, we cannot be sure about the future, but we can refuse to do what we think is likely to produce senselessness between people.

91 Cf. Philip Pettit's *Republicanism: A Theory of Freedom and Government*, New York: Oxford University Press, 1997 and Glenn Coulthard's *Red Skin White Masks: Rejecting the Colonial Politics of Recognition*, Minneapolis: University of Minnesota Press, 2015. On the basic need to relate to the world with intelligence, see especially Daniel Scheinfeld, Karen Haigh, and Sandra Scheinfield, *We Are All Explorers: Learning and Teaching with Reggio Principles in Urban Settings*, New York: Teachers College Press, 2008. See also Martha C. Nussbaum, *Women and Human Development: The Capabilities Approach*, New York: Cambridge University Press, 2000.

92 Hirsch and Norton, 2012.

93 See my 2006, lecture 6 on such a thoughtfulness criterion. See also my 2014.

94 See my 2006, lectures 6–8.

95 See Chapter 2 of this novel as well as Vogel, 2015.

96 See Chapter 3 of this novel on relating through the land across time.

97 That is, my *ēthos*. Here is the *ethical* shaping anthroponomy implies. See Thompson's and my 2012 on a "virtue ethics" approach to planetary-scaled, environmental adaptation – in that book focusing on global warming and, with my essay therein, on the risk of the sixth mass extinction. See also my "The Practice of Ethics," "postscript a" of my 2017a.

98 Cf. Mignolo and Walsh, 2018 on the plurality of worlds as a decolonial attitude.

99 On a critical attitude, see Michel Foucault, "What Is Critique?" in James Schmidt, ed., *What Is Enlightenment? Eighteenth Century Answers and Twentieth Century Questions*, Berkeley: University of California Press, 1996, pp. 382–398.

100 See Young, 2010, also Rancière, 2004.

101 Mignolo and Walsh, 2018.

102 See Young, 2010 and consider again Geoff Mann and Joel Wainwright, *Climate Leviathan: A Political Theory for Our Planetary Future*, Brooklyn, NY: Verso, 2018, especially "Climate X."

103　In my 2016, I called this consequence, the "civic relation." I think it is simpler and more direct to call it a "political" relation.

104　See Chapter 2 of this study re: license and its arbitrarianism.

105　See Young, 2010; Mann and Wainwright, 2018 and cf. Frank Biermann, *Earth System Governance: World Politics in the Anthropocene*, Cambridge, MA: MIT Press, 2014.

106　That is, my attitude implicit in my *ēthos*.

107　While not every decolonial option must be anthroponomic, since decolonial traditions may not advance an explicitly teleological view regarding anthroponomy and its planetary temporal and spatial dimensions, still any *anthroponomic* orientation must be decolonial, and thus decolonizing.

108　See my and Haufe's 2017.

109　As Kyle Powys Whyte demanded of allies, not merely romantic sympathizers; panel on the future of environmental justice, EJ20, University of Sydney, November 2017; see also Ta-Nehisi Coates, "The Case for Reparations," *The Atlantic*, June 2014, accessed September 29, 2019.

110　Mignolo and Walsh, 2018.

111　Relational reason is technically separate, but cooperative with, theoretical and practical reason. See my "Do You Have a Conscience?" *International Journal of Ethical Leadership*, v. 1, 2012, pp. 52–80.

112　Biermann, 2014; Dryzek and Pickering, 2019. Wildcat, 2008 and Whyte, 2018 point to important know-how and practices that ought to shape new directions in governance, but strictly speaking even there the temporal and spatial dimensions of these practices are not yet geological. That, of course, is the result of colonialism's binding of the globe and now hurtling of our current biological order toward ruin. But it is, unjustly, our situation and one that must now become part of anthroponomic governance.

113　See Chapter 5 of this novel.

114　This book is a fiction, and so – some say – is all of reality. But while we can never be morally absolute because of our risk of error or vice, moral relations, when we hold them as authoritative, subvert fictionality. My love for my family is not a fiction. That I must respect you is not either. Our lived moral relations are more than fictions, because they *have* to be, and this book is only a propaedeutic to a certain kind of living. Cf. Conversation with Alex Shaker by WhatsApp, September, 2019.

115　Cf. Ahmed, 2006 on desks and offices as scenes of philosophy deserving queering (swerving away into other spaces).

116　For instance, one might consider organizing with Extinction Rebellion or some such similar group.

117　Cf. Luce Irigaray, *Sharing the World*, New York: Continuum, 2008.

118　Dryzek and Pickering, 2019.

119　Cf. Whyte, 2018; Hirsch and Norton, 2012.

120　As it appears to be in Dryzek and Pickering, 2019.

121　This is an inversion of the infamous Heideggerian formula by which anxiety is "the possibility of possibilities." The inversion is actually Heidegger's modality for death. See Martin Heidegger, 2010.

122　Cf. *1 Corinthians 7:31*, and *Hebrews 13:14*, obviously without theological or specifically Christic content.

123　Cf. Heidegger, 2010 on the ontological as opposed to the ontic.

124　Cf. Neiman, 2008 on unsettled happiness as Odyssean striving. On mood, see Heidegger, 2010. Note that I am arguing for something different than Jonathan Lear's "radical hope," that "hopes against hope" to the possibility of goodness beyond the cataclysm of one's entire world collapsing. See Jonathan Lear, *Radical Hope: Ethics in the Face of Cultural Devastation*, Cambridge, MA: Harvard University Press,

2006 and on "hope against hope," Stanley Cavell, *Emerson's Transcendental Etudes*, edited by David Justin Hodge, Palo Alto: Stanford University Press, 2003, chapter 8. Lear's study, not atypically for his works, has no explicit grasp of moral relations that *are* what pass between collapsing and reconstituting *ethoi*.

125 Richard Burnor and Yvonne Raley, *Ethical Choices: An Introduction to Moral Philosophy with Cases*, New York: Oxford University Press, 2018.
126 Cf. Friedrich Nietzsche, *Thus Spoke Zarathustra: A Book for All and None*, translated by Walter Kaufmann, New York: Modern Library, 1995.
127 Cf. my 2006, lecture 6, "Being True to Ourselves."
128 Cf. Irigaray, 2008.
129 See Chapter 2.

Figure E.1 Belle Valley, Ohio, May 2016

In Belle Valley

Over half a year ago, I had a strange dream in which my mother, who had recently died, was sitting up on the hills of the homeplace of her immigrant grandparents, "Starki" and "Starka" Bendik, in Belle Valley, Ohio, looking southward toward West Virginia.[1] She looked back at me and pointed ahead to "them" – the reference was mysterious.

Belle Valley is today little more than a single road with dilapidated houses and a barely functioning gas station and convenience store. The newest building is an evangelical church up on a hill above the village's single road. The Lutheran church where our immigrant ancestors went was demolished long ago.

Belle Valley, like much of coal country, is desperate. But people live there, and the more than human world is lush even around the environmental destruction and slow violence of mining.

There, too, are ghosts of older nations, indecipherable to a colonialist like myself. Wounded and scarred, the place is still layers of home.

*

The question of what others can make of anthroponomy is a question about finding autonomy-in-relationship within a scarred world. Disagreement here is epiphenomenal compared to the division of wounding. A history of violence is more than a history of disagreements.

The question of how I can support others anthroponomously is a question that is more than cognitive. I think of my family as I ask this question, because of the *relational capacities* – the capacities to have good relationships – that are demanded by the challenge of decolonial work to change the wanton, structurally unjust processes of colonialism and actually existing capitalism. The world is full of ghosts, pathology is part of the very fabric of settler and colonial society, and fear is heavy in the air. We need relationships that are thick enough to be with each other.

They have to be emotionally mature and autonomous. The main way we can support others making something of anthroponomy is through moral accountability with a history of violence that stretches backwards through layers of wounding and scarring and forwards up ahead to "them" who will inherit a haunted and often ambivalent land.[2] To be able to countenance and incorporate such a process demands capaciousness, patience, sensitivity – also endurance, perspective, persistence, and a strong sense of what is right stabilizing interactions as one works through bodies of hurt and loss. These are all thick words for accountability and autonomy.[3]

Relational maturity is the heart of the demand that makes anthroponomy difficult. That may seem odd, given that what seems novel about anthroponomy as a moral form are the planetary dimensions of it. After all, it is novel that *homo sapiens* must learn to think like a planet in order to become sapient. But whereas we are far from developing the ecological reflexivity that we need on planetary scales, it seems to me that the greatest obstacles to doing so are emotional, relational, and moral. We certainly need the know-how to mind the planetary effects of our social processes, but we have a great deal of knowledge concerning them. What seems lacking most conspicuously is the moral accountability needed in our social processes to demand that we involve our knowledge in the construction of practices and relationships that know how to be mindful of planetary, environmental effects. But to develop such accountability demands becoming mature people who can and will relate autonomously with other people as moral equals. It demands thickening in relationships around great indignation, histories of injustice, and trauma.

In Plato's Πολιτεία,[4] the persona "Socrates"[5] develops an argument for the relationship between being just and flourishing as a person through an analogy between the city and the soul. The structure of our psychology and its maturity becomes as central as the structure of our political order and its justice. The inner structures the outer, and the outer structures the inner. Plato's recounting of "Socrates's" argument is instructive for anthroponomy.

Anthroponomy within social processes shapes people – the outer shapes the inner. To live mindful of wantonness even on planetary scales and to rest in disagreement-in-relationship across time and around the planet shapes how one lives as a person. But from the perspective of trying to bring anthroponomy into being, it is the *inside* that seems to be crucial for the development of the outside. Being able to handle disagreement that involves trauma and legitimate, multigenerational outrage and indignity seems crucial to shaping anthroponomy-in-becoming. A kind of environmental maturity seems crucial to the start. Soulfulness is in demand now.

Soulfulness with others and with oneself, linked to the land and thinking like a planet, is challenging. For my society, it must go backward, not forward, into emotionally complex accountability for the past of colonialism that overlays accountability for the trajectory of our planet. There is overtone time in this complexity.[6] The future is in the past, and the past is in the future. Soulfulness layers time.

In a world structured by domination, erasure, and deflection,[7] accountability opens sense-filled possibilities again. This is part of what makes its processes so painful. To live in domination, erasure, and deflection is to live resigned to a world where what is isn't what ought to be and where one's basic capacities as a person – one's moral judgment, intelligence, and sincerity – are denied and repressed. When people and processes become accountable in such scenes of injustice, it is as if the sky had cracked open momentarily from out of a leaden, unending storm. It is destabilizing, unnerving, and makes one suddenly vulnerable again from out of protective invulnerability and disassociation. It shakes open possibility and also tempts reactions. Would it be easier to not have hope or the pain of feeling that there could be justice? Do we know how to live in a world that makes sense between people?

Inside ourselves, there has to be a psychological transfer from being invested in power over people – from protective control born out of perpetual insecurity – to finding one's bearing as a person in being accountable. Getting things right between people has to become the mark of one's time in society, far from the insecure and egotistical psychological formations of profit-seeking, status-lording, or objectification of others for the sake of controlling one's environment. This is a big "ask" from people when the entire society is structured by insecurity and wrong. It's to ask for integrity in a situation that isn't trustworthy. Yet this is the moral thing to do. It is morally required of all of us.

*

Three years ago, I came through Belle Valley for the first time as an adult. The green was lush and rolled away from the roads. There was mist, and the grass was wet from morning rain. I heard bird calls and the occasional cricket – and then the sound of the freeway far off slicing through the hills down into the coal country of West Virginia. In the midst were still poor homes, hard-driven workers, unemployment, opioids, and the despair that politics and economy make little sense between people.

Given the quiet, I remembered the feeling of the plain character of my mother's side of the family. If there was one thing we were supposed to be, it was right in our relations. If you do not have your character, what do you have? If your word isn't good, what good are you? If you don't have integrity, what kind of structure do you have for this rough life? These were challenges, not compliments. I did not get the sense from my family that we ever complete the work these questions involve.

One person who worked hard on her relations her whole life was my mother. I remembered how Esther Ann Bendik, who was a singer, was open in the universe of life. Her auditory sensibility reverberated with the living sounds of land. I remembered how she was in moments when she listened, when the wide surround of life carried through her like electricity conducted through the sky.

Notes

1 See Chapter 3 of this fiction.
2 Regarding "them," see the dream near the end of Chapter 3.

3 On thick verses thin moral language, see Philippa Foot, *Virtues and Vices*, New York: Oxford University Press, 1978. Thin moral language is abstract and lacks immediate connotative power in guiding our judgments. Thick language has, by contrast, strong connotative power. "Right" is thin. We don't know what it involves too easily. "Courageous" is thicker. We probably have many *people* about whom we think in relation to the word.

4 Plato, *Republic*, translated by G.M.A. Grube and revised by C.D.C. Reeve, 2nd ed., Indianapolis: Hackett Publishing, 1992.

5 Plato's Socrates was a fiction based in memory and turned into a myth.

6 On overtone time, see Chapter 3.

7 On domination, see Glen Coulthard, *Red Skin, White Masks: Rejecting the Colonial Politics of Recognition*, Minneapolis: University of Minnesota Press, 2014; on erasure, see Kyle Powys Whyte, "Settler Colonialism, Ecology, & Environmental Injustice," *Environment & Society*, v. 9, 2018, pp. 129–144; and on deflection, see Chapters 1 and 5 of this novel concerning the misdirected responsibility implied by the common meaning of the name "Anthropocene" – deflecting responsibility away from specific social processes onto humankind as a whole.

On the Farm

Julia D. Gibson

The first time I encountered Jeremy Bendik-Keymer's concept of anthroponomy was at the annual meeting of the International Society of Environmental Ethics (ISEE) in 2019. In fact, our papers shared a session, somewhat humorously (if accurately) titled "Issues RE: The Anthropocene." *Dearest Anthropocene, you've got 99 problems but a golden spike ain't one. xoxo Jeremy and Julia.* Both the project and its creative written form resonated with me and the research I was at ISEE to present, my own response to wantonness – climate justice for the dead and the dying. Jeremy must have felt something similar to have invited me to disrupt/respond to the rich philosophical work presented in these essays and his novel as a whole.

For my part, I was drawn to the project as a sincere attempt on behalf of a fellow settler scholar of environmental philosophy to grapple seriously with the reality of living and working on stolen land in the time of climate change. Anthroponomy unfolds within scarred worlds not imperiled wilderness. It was also vital to me that anthroponomy was being developed in direct opposition to colonization, both past and ongoing, and its offshoot – the Anthropocene hypothesis. All the better that anthroponomy delved into relationality and temporality. My own contribution to ISEE had much to say about mainstream environmentalism's intergenerational and temporal shortcomings and how they contribute to the neglect of the unjustly dead/dying and transformative futurities.

But I'm not here to rehash that project, at least not directly. Nor am I terribly interested in tracing every last philosophical thread in the complex tapestry of anthroponomy whether to map or unravel them. No, what I want to do here is take seriously the question that the novel leaves us with – what can others make of anthroponomy? How can this work support others' and be enriched (or challenged) in the process? And because the journey that produced this closing query was both deeply personal and very intentionally located, I do not approach anthroponomy as a scholar removed from their analysis but as engaged philosopher on the land I call home. Like the author, I would "come to terms with where I am and the specific location of the life I've inherited."[1]

*****

When I reflect upon land, I think first of my family's farm in the Lower Hudson Valley. This is the place and the relations that have inspired so much of my work

as an environmental philosopher and to where I could not help but be transported –
in both mind and body – while immersing myself in these essays. Indeed, it was
on an evening spent cocooned in the fall dark of the Corn Crib and the familiar
emotional tumult of planning for the farm's uncertain future that I realized it was
here that anthroponomy could mean the most to me. And not just because this is
where my relations (landed, familial, or otherwise) are the thickest, but because
I have been avoiding the full implications of what decolonization and climate justice should look like on/for this land. It's time for me to do that work and I want to
know how anthroponomy can help, as well as where it may fall short.

There are many stories that you could tell about Ryder Farm. Here is the one
I grew up with. I am Julia D. Gibson, daughter of Henry Hall Gibson, son of Kath-
arine Belden Ryder, daughter of Ely Morgan Talcott Ryder, son of Henry Clay
Ryder, son of Colonel Stephen Ryder, son of Eleazer Ryder, who built the Ryder
family homestead–The Sycamores – in 1795 on the crest of a hill in the town of
Southeast (now Brewster) in Putnam County, New York. The Farm was incorporated in the early twentieth century to safeguard Aunt Mary in her dotage, result-
ing in the unusual arrangement wherein the fourth of July family reunion begins
with an annual shareholders' meeting. Gifted a share by my grandmother as a
small child, I was raised to regard these meetings with great solemnity. Though
often quite contentious – with the family sometimes splintering along ideological
factions or bloodlines – the message was clear; this was our Farm and, as Ryders,
we were charged with caring for and protecting its future while preserving its past
and the labors/wishes of Ryders who come before us. To fail to do so would be
devastating beyond comprehension.

And the anxiety of losing the Farm was seemingly grounded in reality, because
both Farm and homestead were understood to be perennially under threat by
some outside – and occasionally inside – force, whether it be skyrocketing taxes,
encroaching development, changes to tradition, falling tree limbs, or a younger
generation that carelessly slammed screen doors and left hammocks out in the
rain. My own grandfather was represented in this tumultuous story as one of the
Farm's saviors. He who married into the family, fell in love with the Farm, and
retired here in the late seventies to found one of the first organic farms in the
Northeast and secure an agricultural tax abatement. Before him it was my great
grandfather Ely; after, my cousins Betsey and, most recently, Emily, who have
sought to cultivate art and music alongside organic produce.

Caretakers abounded in the story as well. Women, usually, and men who knew
every nook and cranny of the Sycamores, every berry patch (and when they'd
ripen), the lineage and stories of all the cousins, worn paths through the fields
down to hidden lake beaches, and how to make themselves known and heard
in their own fashion. Aunts, uncles, and cousins who rang the dinner bell, who
tirelessly maintained the structures, whose names passed on to rooms, dwellings,
jams, and children (myself included), and who loved the Farm and by so doing
saved it over and over again in the everyday way. In the story I was told (and lived)
someone was always saving the Farm. It was a place worth saving. Someone was
always caring for the Farm. This was a place – a family – deserving of care.

As you may suspect, plot holes abound. For one, the land's story prior to the Ryders – *whose home was this before Eleazer built his sycamore shaded homestead on the hill?* – is entirely absent. Neither are the stories of family and Farm much contextualized within the larger political tapestry of those communities they have long belonged to beyond the grumblings about the influx of New York City folk to Putnam County and their impact on the tax rates. Instead, our story and the Farm's have rolled (and been steered) along mostly untethered to the larger stories to which they belong, generating temporal and spatial insulation from the outside world(s). The feeling of stepping into a place "out of time" has become part of the Farm's appeal. We Ryders remember much, but we do not always remember well.

It is unclear how much shared sense the Ryder narrative makes, except perhaps to fellow settlers seeking innocence[2] and escape. Our selective memory prevents us from being held accountable to those written out of the Farm's story, however democratic and conflict-rich our internal workings. How might the narrative be reworked to make room for the kind of disagreement that anthroponomy demands? How can it help make sense of our responsibilities under the full weight of climate change as environmental injustice? This seems a lot to ask of a narrative, but – as others[3] have noted – the kinds of stories we tell about ourselves in the time of climate change matter deeply. So, here is another story you could tell about Ryder Farm, one that I think anthroponomy can work with.

I am the four times great grandchild of Eleazer Ryder, who leased this land following forceful colonization and warfare both of which drove the original caretakers of this grove of sycamores, rocks and fields, forest, swamp, and lake – the Wappinger and the Munsee Lenape – North, West, and South. Ryders have farmed these 127 acres for close to 225 years in ways both violent and loving, thoughtful and thoughtless. In recent years, organic farming has been a way for Farm and family to grow our inter- and intra-species partnerships and reconnect to the communities and ecologies that sustain this place (and us) even as weather patterns and worlds shift. Art and music have long flourished here, but they do not erase our mistakes. As long as the story of our past/present is incomplete, the Farm's future will be fraught. My children's children may not own this land, but they will always be tied to this place and know the responsibility of (at)tending to those who once, do, and will call(ed) it home, especially those who did/do not leave on their own terms. Loving[4] the Farm is the work that makes us family.

＊＊＊＊＊

At ISEE and elsewhere,[5] I critiqued mainstream environmentalism as problematically past-oriented, that is, temporally and normatively constrained to strategies, tools, and ethics geared toward pausing or winding back the ecological clock at the expense of the victims and survivors of environmental injustice. *Make the environment great again!* Climate change has undermined the practicality of past-oriented environmentalism but not, as yet, the desirability of its carefully curated Past. Across the Anthropocenic time-/land-scape, environmental apocalypse is firmly rooted in "humanity's" past but it does not live there. Here, the World has not yet shattered, though the cracking of ice is audible.

Suffice to say, it is deeply important to me that any environmental ethic I plant on the Farm does not depend upon or reinforce this past-oriented paradigm or its temporal-normative underpinnings. Ryder Farm already has its own idealized past and, relatedly, quasi-apocalyptic narrative to contend with. It doesn't need another. So, I must ask: does anthroponomy avoid the pitfalls of past-oriented environmentalism? *Spoilers*. The short answer is yes.

Anthroponomy grows out of and is explicitly designed to directly confront colonial violence, industrial destruction, and capitalist exploitation. From the author's time- and place-stamps to the underlying theory, scarred worlds and their ghosts are front and center throughout the novel. Indeed, the standard formulation of the Anthropocene is heavily critiqued, in part, for the elements of the past that it works so hard to hide. But anthroponomy offers more than a refusal of a "nostalgic and idealized past,"[6] for the pursuit of anthroponomy requires nonlinear understandings of time. Standard Anthropocene temporalities cannot accommodate the kinds of intergenerational ethics and land-based politics that are called for by this approach.

And, so, we are introduced to overtone time. Inspired by Indigenous temporalities (e.g., spiral time[7]) and a family tradition of song, overtone time enables anthroponomy to embrace complex temporal harmonics. Doing environmental ethics – or simply living – in overtone time means attending to the past, present, and future as layers working in multidirectional relation to each other rather than (or in addition to) as touchstones in a unidirectional current. *If time were water, it would not be a river but the whole damn hydrosphere.* But whereas song travels through air, temporal relations resonate through the land.

This is not to say that anthroponomy is untethered from the realities of climate change as an intergenerational injustice. All the way back in the first essay we are told "it is strictly impossible for [present and future] generations to conduct reciprocal relationships" given causal asymmetry, i.e., that generations can only affect those who come after them.[8] In fact, the author goes so far as to claim that we are "predominated over" by past generations.[9] Perhaps I am still too enmeshed in linear temporal discourse – *this strikes me as a genuine possibility* – but some degree of causal/temporal asymmetry seems undeniable. Yet, I struggle to reconcile the idea of generational predomination with the richness of overtone time and the ethics that develop alongside it in the latter half of the novel.

For overtone temporality is robustly intergenerational in a way that, upon my reading, is in tension with the impossibility of reciprocity. In overtone time, generations (dis)harmonize through layered moral relations of care. This dynamic often gets couched in terms of "lineages of care" stretching forward; one generation caring for the next and so on.[10] But holding ourselves accountable to our lineage involves not only descendants but ancestors.[11] In the present, we care for both past and future generations. Through the land we are held accountable to both. In fact, the intergenerational care described and enacted in/by chapter three is directed "backwards" as much as "forwards." Attentive to ghosts, lost loved ones, and future persons, anthroponomy works to maintain intergenerational *community*. When these communities function well, what we have are *reciprocal relationships* of care between generations living, deceased, and those-yet-to-be.

This is certainly how it felt growing up as a Ryder on the Farm. I recently caught a *Radiolab* broadcast on my drive down to a family meeting.[12] The episode was almost a decade old and concerned cities, but it took an all too poignant turn when it posed the question of why people can have such a hard time letting go of a place, even one that's dying. The answer: it's almost unbearable to have to give up the experience a life lived on a multigenerational plane. It took me a long time to realize that what I/we have on the Farm is special in this regard. It took me longer to understand that what we have here was/is only made possible through the rupturing of a deeper multigenerational plane.

Us Ryders too are a songful family, though more prone to rounds than layered harmonies. *Come follow, follow, follow.* Overtone harmonics are certainly present, but the cyclical and indeterminate quality of rounds add further layers and chaotic goods. When singing rounds (with Ryders, at least) it is never entirely certain who will join in, for how long, or whose voice will carry the last note. Last-in rarely equates to last-out. *Wither shall I follow, follow thee?* Intergenerational flexibility and indeterminacy are essential for the kind of temporality needed and, to a certain extent, already growing on the Farm. Rather than narrowly apocalyptic framings that tend to make us conservative in our histories/futurities and reactive in our decision making, I would have us cultivate expansive intergenerational thinking. So, moving forward, I choose the suggestive intergenerational reciprocity that unfolds around anthroponomy as it roots itself ever more firmly to/in the land. *To the greenwood, greenwood tree!*

*****

Before working with the full regulative implications of anthroponomy, it will be helpful to get a better sense for the types of autonomy practiced on the Farm. What kind of community do we have here? Who is involved and how? As a relational ethicist, I love that community is at the heart of the kind of autonomy that goes into anthroponomy. Moreover, it is refreshing to see autonomy explicitly working against "liberty, license, and capitalism" – not to mention colonialism – within mainstream (adjacent) environmental discourse.[13] If ever there was an autonomy I could see getting cozy with environmental justice, this is it.[14]

The "who" of Ryder Farm begins with the s/Sycamores and builds outwards. The question then becomes whether to start with the homestead or the trees. If we build our sense of community from the Sycamores as a homestead, the Ryders get situated at the center. A farming family with longstanding ties to established settler families and local banks, the Ryders became prosperous enough to hire others to plow and pasture until financial support dwindled and taxes shot up in the mid-twentieth century. The communal ties – e.g., to the Putnam County Land Trust, Southeast Museum, Northeast Organic Farming Association (NOFA), and Hudson Clearwater project – to grow out of this financial crisis emphasized land in concert with family and local heritage. Without the financial means to remain detached, we have had to become more accountable to (some of) our neighbors. These (relatively) more recent associations were leveraged when the family sought the matching funds needed to apply to the state of New York for the sale of development rights ten years ago. Though our application was ultimately unsuccessful, we were informed that the level of community buy-in the Farm garnered had been far and above our competitors. Ryder Farm makes sense to/ with more than just Ryders. And with the introduction of SPACE on Ryder farm, a thriving artist's retreat,[15] Ryder Farm's community grows in numbers and across vectors that we couldn't have imagined a decade ago.

If, on the other hand, we start with the sycamores,[16] which I believe is not only something that anthroponomy can accommodate but would encourage, then who we understand as the community of Ryder Farm looks somewhat different. The first thing I learned about sycamore trees growing up was that they know and signal where groundwater can be found. That's why the Sycamores was named for them and built within their grove. From the sycamores we began to make sense of this land – above and below – and ourselves through it. Over the past two hundred and twenty-five years, Ryders have become increasingly intimate with this place: field, fen, and forest, bobcat to berry, nut to worm. Our community expanded through the efforts of those Ryders who took the time to cultivate and maintain relationships with our nonhuman cohabitants and neighbors, to know the sycamores and not just the Sycamores. Grown more from individual initiative/ inclination than familial structures or expectations, such relationships have admittedly waxed and waned. After years of only supporting a tenant dry-herd and halfheartedly growing hay, my grandfather's decision in the seventies to become involved with the organic movement and launch a diversified produce operation revitalized both the family's connection to farming and our sense of the land as reciprocally nourishing.

My own recent work – in conversation with the board, our farmers, and SPACE – to articulate an animal policy for the Farm is in the tradition of this second sort of sycamores sense-making and community-building. *Sycamores countersense?* The policy begins with our overall vision:

> We recognize that a farm is inherently an interspecies community. Whether domesticated, wild, or liminal, non-human animals have been and always will be integral members of the Ryder Farm. As the caretakers of this land, we commit ourselves to living together well with our animal cohabitants. Striving to do so is a vital part of what it means to tend to and balance, as best we can, the familial, ecological, financial, historical, and spiritual well-being of The Farm.[17]

In a footnote, we explain what we mean by "living together well":

> Roughly speaking, this phrase is meant to capture the acknowledgement that we have ongoing relationships with the animals we live, work, and play with, relationships of the sort that generate obligations of reciprocity, compassion, and respect. This document outlines certain strategies and ideals for meeting these obligations, but it recognizes two important realities: (i) our relationships with different individuals and communities of animals, as with the members of our own species, do not have a single mold and (ii) reasonable, kind people can and do disagree about what these obligations are and how

best to meet them. Thus, this policy is intended as firm yet flexible framework that will provide RFI and its partners with both the structure and the adaptiveness needed to properly attend to our animal partners. It should be a living document that facilitates, rather than stifles, dialogue and is amended and added to as needed.[18]

From the built-in disagreement to the undercurrents of relational reason running throughout, this approach exemplifies the kind of autonomy/community that makes anthroponomy possible.

But as with anthroponomy, there is a sense in which what we're doing here is spinning our own sort of speculative fiction. Ryder Farm needed a formalized animal policy because of a longstanding pattern of conflict and, at times, neglect, with regards to domesticated animals in particular. In order to know what we can really make of anthroponomy on the Farm, we need to be honest about the type of community – and, thus, autonomy – we've actually got. A tradition of sycamores-sense-making, however subversive, does not erase the homestead mentality that makes Ryders so prone to regard the Farm as a business/investment first and foremost and ourselves as landlords. This is practical reason at its finest and results in more than animal neglect. In addition to the sycamores' original human partners, the Wappinger and the Munsee Lenape, Ryders as a whole overlook our connections to the robust local Guatemalan community, who have helped us harvest and prepare for market time and time again. We cannot attempt to make sense with or hold ourselves accountable to these communities, among others, if we refuse to acknowledge them as part of the Farm and the larger/longer story of this land we call home.

Recent efforts to mobilize against local anti-immigrant politics and SPACE's sliding scale CSA are promising examples of community building/accountability. But I take the author seriously when he writes that relational maturity requires emotional maturity. There are internal cultural and structural changes that will need to happen if Ryders are to be "able to handle disagreement that involves trauma and legitimate, multigenerational outrage and indignity."[19] In addition to reworking the story of the Farm, it is vital that Ryders cultivate trustworthy communication practices. For the homestead narrative has affected not just who we take ourselves to be but how we grapple with disagreement. Reaching out to those we have displaced, erased, neglected, and excluded is irresponsible if we are not also working to build the skills and structures necessary to hold ourselves accountable to them when things get hard.

Trust me when I tell you that Ryders are no strangers to disagreement. We are seventy-five shareholders and at least twice as many family members across three ancestral branches. I won't pretend that everyone is equally engaged, but most of us love the Farm dearly and many (myself included) consider it home. Needless to say, we don't always agree about how to run this place. *Ever live-tweet a family board meeting?* On the surface level, we use Robert's Rules of Order and democratic voting procedures, but these tell only part of the story. Every person who has invested themselves seriously in the running of the Farm has been wounded deeply at least once. Most of these wounds are never properly tended. Allowed to fester or scar over, they become part of the emotional landscape of the Farm where they operate as significant obstacles to mutual accountability and trust.

I believe this dynamic is due in large part to Ryders' inability – collectively and individually – to acknowledge the fullness of what it means to love this land and to sit with the difficult feelings that arise when this moral/emotional relationship feels threatened. Often, this manifests as practical reason – in the form of the financial bottom line – being wielded to discount relational reason and keep it from getting too firm a foothold on the Farm. We can value the land non-instrumentally as long as this does not diminish its "real" value. Other times the reminder that the Farm is a business operates as a smokescreen for reasons/goals that have nothing to do with practical reason. Sometimes this is strategic. Usually, however, it is an unintentional consequence of systematically devaluing the relational mode; the right tools don't feel legitimate or within reach, so we settle for the practical.

But it is also here – beyond the practical – that us Ryders are at our most vulnerable. So, we use smokescreens and don't dare dig too deeply. We "put up" with disagreement and each other rather than risk further wounds.[20] Homestead sense-making – as well as the other liberal community norms Ryders have internalized – make us feel like this is the best we can hope for. As the author's exploration of anthroponomy so deftly illustrates, however, in this mode of autonomy we not only fail those (humans and nonhumans) whom we exclude from our communities but our*selves*. For Ryders to become fully accountable to each other on, through, and to the Farm, we will need to figure out how to adapt sycamores sense-making to hold ourselves open and vulnerable to internal disagreement without wounding each other. Just as we are taught to navigate the raspberry thickets and black walnuts on the bowling green, we can learn to make better sense of/with of each other without injury. And we must, for if Ryders cannot hone the skills to fully work through our own disagreements, I don't like our chances of practicing anthroponomy with those we have wronged.

*****

Given who/what we take Ryder Farm to be and how we operate, what can we say about Ryders' relationship(s) to the land? *Cut to the chase*. Are we wanton? As a Ryder, this is a difficult question for me to answer, especially given what I know to be at stake in answering it – the author makes clear that wanton communities have no rights to land. I cannot deny there is a deep part of me that strenuously rebels against the idea of Ryder Farm no longer existing, not only for me but for my descendants as well. I did, however, just make a particular point of the necessity of embracing the vulnerability that comes with loving this place. Thus, I need to leash that impulse, even if it cannot be quashed. Moreover, there's a wiser part of me that knows rescuing settler futurity is not the point.[21] Ryders need to get used to doing right on/by the land without fixating upon where that leaves us. And, yet, even knowing this I cannot answer the question of wantonness completely, not because I do not have the proper distance – that is actually useful[22] – but because there are other kinds of expertise and proximate perspectives that will need to weigh in. Still, I can offer my piece of the picture as a Ryder trained formally and informally in environmental ethics and interspecies politics.

On the face of it, it seems absurd to ask whether a settler farm purchased in 1795 and anchored by a homestead is wanton. "A specific locale of humankind produced wantonness" and us Ryders are undeniably part of this geography.[23] A better question, then, would be if/how we are *not* wanton? Here, in addition to the framework of relational reason, I have in mind Robin Kimmerer's discussion of becoming "naturalized" to place.[24] Illustrating her point through the divergent strategies used by the Turtle Island transplants kudzu and common plantain (also known as White Man's Footstep), Kimmerer writes,

> Maybe the task assigned to Second Man is to unlearn the model of kudzu, and follow the teachings of White Man's Footstep, to strive to become naturalized to place, to throw off the mind-set of the immigrant. Being naturalized to place means to live as if this is the land that feeds you, as if these are the streams from which you drink, that build your body and fill your spirit. To become naturalized is to know that your ancestors lie in this ground. Here you will give your gifts and meet your responsibilities. To become naturalized is to live as if your children's future matters, to take care of the land as if our lives and the lives of all our relatives depend on it. Because they do.[25]

Thus, another way to frame the question I'm after here would be this: Are the Ryders more like kudzu or plantain? Are we good neighbors working toward reciprocity or have we made our home at the expense of others?

I cannot claim that the Ryders as a whole have become naturalized to this place. There is a distinct resemblance between the way kudzu operates and the ecological practices (e.g., rodenticide) informed by homestead, landlord, and agribusiness mentalities and relational norms. But there are those Ryders, such as the caretakers in my story, who aspire and, at times, succeed in becoming more like plantain. *Belle, the berries miss you.* As a family, the Ryders' commitments to organic farming, nonhuman flourishing, and the arts generates policies and practices that often – though not always – run counter to kudzu-like forms of life. On both the communal and individual level, when Ryders embody sycamores countersense they move the Farm closer to a naturalized interspecies community. In these contexts, relational reason is not absent, but it does have distinct competition. Moreover, as with the s/Sycamores themselves, relational and practical reason on the Farm are sometimes coextensive. For example, the Ryder practice of reciting our lineage is both a WASPy purity ritual rife with colonial-erasure *and* a mechanism for locating individual Ryders relationally – *how are we related again?* – and as discrete ancestors in place/on land.

Admittedly, even the countersense of sycamores has its "defects" as far as relational reason is concerned.[26] Celebrating the sycamores as water-finders but leaving out who tended/loved these trees and the water beneath them prior to the Ryders does more than compromise the history of the Farm; it renders our community in the present incomplete and complicit with colonialism. Unlike Sycamores sense-making, however, as a fundamentally relational approach working against land abstraction, sycamores sense-making is compatible with anthroponomy and, thus, (potentially) with naturalization and decolonization. Growing up with the contrast – both stark and subtle – between the Farm's failures and its departures from settler-colonial logics has shaped the environmental philosopher I have become. This was where I started to learn, however incompletely, about other ways of relating to land. As I learn more, I feel increasingly beholden to this place, to doing right by and through it. The ancestral weight of the Farm is an incredible gift, an undeserved privilege, and a daunting responsibility. I take some small comfort in the thought that the Ryders are perhaps not totally without homegrown tools. Across nine generations, we have gleaned something of what it means to work on/for/with this land, though we still have much to learn and to do if we hope to one day be naturalized. Perhaps anthroponomy can help.

*****

"Here you will give your gifts and meet your responsibilities."[27] What are the Ryders' responsibilities to and through the Farm under settler-colonialism, capitalism, and industrialism? In the time of climate change? The author offers anthroponomy as a strategy or "regulative idea" for attending to both such questions.[28] Though we may not agree on their exact constellation, for the purposes of this essay it is enough that these phenomena are integrally connected, with land abstraction playing a major role across them all.[29] For Ryders do not need to know the precise role that land abstraction plays in colonialism or climate change to weed out this logic on the Farm and to work against it elsewhere. However, moving forward we will need to be on guard for how shoring up sycamores countersense – even as a relational mode of autonomy and sense-making – has the potential to lend itself to settler moves to innocence.[30] *My ancestors have sacrificed to preserve this place. We know every rock and tree and creature (has a name).* Not all decolonial projects must be anthroponomic but working toward anthroponomy must always be a decolonial/decolonizing process.[31]

Anthroponomy "begins with disagreement over the erasure of colonialism right here."[32] On Ryder Farm this means not only doing the research to correct our narrative but cultivating reciprocal relationships – if desired – with the descendants of the Wappinger and the Munsee Lenape. Additionally, Ryders should reach out to Indigenous peoples who have more recently made the Lower Hudson Valley their home, for example, the Ramapough Lenape Nation. *If* any of these communities are open to receiving Ryders as partners, the family will need to be especially vigilant and proactive to ensure that its members do not to perpetuate extractive/oppressive interpersonal or communal dynamics, for example, relationships wherein labor or resources only flow one way or in which Ryders react defensively to being confronted with the full implications of the Farm's settler-colonial legacy. Both relational and emotional maturity – and the communication skillsets that go along with them – are necessary for this work. Again, that Ryders commit ourselves to learning how to better handle internal disagreement will be essential.

Guided by anthroponomy, the point of expanding our community to include Indigenous peoples is to – finally – hold ourselves accountable to those whose domination was/is necessary for Ryder Farm to exist, not just in the past but in the present as well. Indeed, making ourselves accountable to these communities means both acknowledging how the Farm came to be and refusing Ryder "sovereignty" moving forward. Anthroponomy is "built on the priority of disagreement," requiring that we hold ourselves open to making sense of the world differently and, thus, to who we may become.[33] Ryders must accept that forging new partnerships may change us and the Farm, however uncomfortable this uncertainty may make us feel. Fortunately, several decades of financial woes mean that Ryders are already familiar with uncertainty when it comes to the Farm's future. *Never thought there'd be a silver lining to that particular kettle of fish.* With some work, our relationship to this anxiety could mirror the author's own journey in this novel: from apathy and dread to excited anticipation.

Anthroponomy is also concerned with overhauling our relationships to/with the nonhuman subjects of colonialism or, in other words, the Farm itself. This land is saturated with moral relations that sycamores countersense has only begun to explore and make sense of/with. In addition to holding ourselves accountable to the Indigenous humans of this area past and present, decolonizing the Farm through anthroponomy would require Ryders to cultivate good relationships with the other living beings who make their home here as well as the ecological reflexivity to maintain them. Our new animal policy is a good start, but it does not address all our salient bonds on this land. Nor does the policy explicitly challenge – *implicitly? perhaps* – our corporate structure or other significant reifications of practical reasoning and colonial logics on the Farm. Conservation easement or non-profit status, for example, may be options worth exploring,[34] however, they must be considered contextually. If (one of) the goal(s) is to maximize familial accountability and involvement, then giving up the democratic procedures that come with having family members double as shareholders in favor of a small non-profit board may be undesirable. Similarly, conservation easement *could* limit us insofar as how we are able to hold ourselves accountable to Indigenous communities. Ryders working toward environmental justice and decolonization should expect to encounter incommensurability.[35]

But grappling with localized complexities is not all anthroponomy would have us do. For this regulative idea also demands that Ryders think and act across larger/ deeper spatial and temporal scales toward the goal of coordinating with others (i.e., humans) to become "collectively accountable for our planetary situation."[36] Embracing these responsibilities is how we participate in the counter-sense of the Anthropocene. While Ryders have considerable, if narrow, practice holding ourselves accountable to past and future generations, we tend not to be as savvy about understanding family and Farm as belonging or connected to larger communities, institutions, and structures. Becoming part of the organic movement and home to an artists' retreat have both helped expand our sense of what the Farm is about. Feeling the early impacts of climate change on our crops has sparked a different sort of revelation about how we fit into the bigger ecological picture. More than ever, Ryders are realizing that the Farm does not exist in a bubble; we will need to build community across towns, counties, states, countries, and generations if this place is to survive as something our ancestors (and others') would recognize as home to the sycamores.

Ultimately, anthroponomy would have Ryders recognize ourselves and the Farm not only as vulnerable to climate change but as drivers of climate injustice, colonization, and capitalism. Committing to non-domination will mean overhauling many aspects of how we relate to the land, to each other, and to other communities of humans and nonhumans across space and time. Living up to anthroponomy is an ongoing project. "A life of responsibility, autonomy, and of non-domination 'is not something you find; it is something you make.'"[37] In particular, anthroponomy would have Ryders consider the colonial legacy of the Farm and whether we are worthy of – let alone entitled to – continued habitation and/or stewardship. "No nation has a right to the land if it isn't thoughtful with it."[38] The question becomes: how many second chances do the Ryders get? *How many have we already wasted?*

*****

As much potential as there is for anthroponomy on the Farm, I do wonder whether it is enough. While anthroponomy never pretends to be the solution to all the worlds' woes, it is put forth as a strategy for working toward decolonial and environmental justice on local and planetary scales across generations. Thus, anthroponomy should work on the Farm, though it will look different here than it will anywhere else. This is as it should be. Though anthroponomy aims for structural change and requires unprecedented coordination, it is fundamentally about relationships-in-place and is mediated through particular lands. Whether anthroponomy is enough to address colonization and ecological harms/wrongs for particular places is just as important as whether it holds together abstractly, a question which I largely leave to others. Indeed, the local level is often the hardest for philosophy to get right.

Some of my qualms have to do with the ideological baggage that anthroponomy brings along for the ride. I footnoted Kant earlier; though his influence is not insignificant, I am mostly mollified knowing that Kant would be rolling in his grave if word ever reached him of what anthroponomy has become under the author's care. The specter of the Anthropocene hypothesis is more troubling. I just don't know if what the farm needs is more humanity, even if framed through the lens of responsibility and coordinated social change/action. Assuming that we could successfully purge the dominant sense of the Anthropocene from the Farm's discourse, would I advocate for Ryder accountability to those we have dominated, excluded, and neglected in terms of species agency? With regards to intraspecies injustice, my (and others') work on the Farm has most often involved highlighting the Ryders as settlers or white Americans rather than as humans. When it comes to interspecies or environmental injustice, I seek to draw out the longstanding nature of our relationships with particular nonhumans, relevant ecological dynamics, the richness of the nonhuman lives/beings in question, or Ryders' roles as caretakers, i.e., elements of relational reason. Emphasizing our humanity in these contexts tends to magnify distance and, thus, fuel practical reasoning and land abstraction. It is not that 'humanity' is altogether inaccurate (or species membership irrelevant), but it tends to be too blunt a tool for politics on the Farm.

Another aspect of anthroponomy's (inter)species politics that may not be ideally suited for Ryder Farm is that this orientation seems to rule out relating to nonhumans and the land as persons/selves. While the author is adamant that we can relate personally – and, thus, morally – to the more than human world, there are times when these relationships – like those with distant ancestors and descendants – come across as "imaginative" if not "deluded."[39] At the very least, we are told that the personification of nonpersons has to traverse deeper differences than it does when relating to other persons. I am not convinced that anthroponomy requires such a distinction; if anything, demanding coordination across a wider range of persons (defined simply as those entities with whom we have moral relations) seems an excellent way to undermine wantonness. Moreover, I could see the personal/ impersonal(-but-personified) distinction backfiring on the Farm, where self-serving interspecies hierarchies are eager to form. I would rather have us overshoot and be too liberal with personhood than be too cautious. Expanding personhood in this way takes nothing away from us Ryders except undeserved privileges. Similarly, it is important for Ryders to be able to relate to this *land* and not just to the many beings who make their homes here.[40] Indeed, through sycamores countersense, I believe (in some ways) we already do. The land of Ryder Farm may be saturated with life, but "it" is more than a sponge or a conduit. The water beneath the sycamore roots is as much a Ryder relation as the trees who showed us where to look. *Come follow, follow, follow.* As articulated in these essays, I am not sure that the 'with' of anthroponomy's relational land logic can capture these persons/relations.

Furthermore, despite its best efforts, I believe anthroponomy runs the risk of encouraging Ryders to think too grandly or abstractly about how to hold ourselves accountable on/through/to/with the Farm. Some may take issue with anthroponomy for adopting a planetary framework that is (at least) in tension with certain worldviews and epistemologies, i.e., totalizing in the third sense. Postcolonial planetarity itself may be at odds with some Indigenous world views if read as taking for granted a common planet.[41] For the Farm, one troubling remnant of abstraction concerns decolonization. There is no doubt that anthroponomy is about more than metaphorical decolonization. From the beginning: "[Decolonial responsibilities] are not abstract but demand relations where we live."[42] And, yet, it remains somewhat ambiguous as to whether/how the repatriation of land – what many[43] consider to be the heart of decolonization – fits into the politics of anthroponomy. Colonization is about more than conceptually separating people from their lands. The author knows this, but sometimes the emphasis on land abstraction skews, ironically, toward the abstract. I appreciate that the settler creation of anthroponomy is understood as an open-ended process that doesn't presume to dictate overly specific decolonial mandates to Indigenous peoples, but I believe it leaves settler/Ryder futurity on the Farm too comfortable, if not particularly secure. Colonial nations should lose their moral right to land, but does this mean that families like the Ryders – both violent and thoughtful, invested in cultivating relational reason – forfeit their land in deed or otherwise? Again, how many chances does anthroponomy give us to turn things around and to grow into our thoughtful ways? Why do we deserve any?

Perhaps the allochthony condition affords no one second chances.[44] After all:

> we shouldn't assume that land must belong to one culture or that the meanings of ancestral lands are the only worlds of this land. The land is larger, more planetary and immemorial than that. The land is a stream of life and of death of myriad forms of life and it exceeds us with the same mystery as the

cosmos. Relatability must be restored in and with this land as a first step to anthroponomy.[45]

What if the descendants of the Wappinger and Munsee Lenape are not actively "practicing land stewardship and seeking a non-extractive, sustainable economy"?[46] Would Ryders no longer be accountable to them on/through the Farm? I understand the requirement that decolonization center land repatriation to be in tension with this orientation. As such, the author's claim here represents a potentially dangerous abstraction away from the heart of what is unjust, violent, and wrong about colonialism as well as what is owed. *Multigenerational planes have been sundered but they are not gone. They are waiting.* At the very least, this passage helps demonstrate one of the ways that the ecological justice envisioned by anthroponomy could find itself incommensurable with decolonization. When it comes to decolonizing the Farm, then, it seems likely that Ryders will need more than anthroponomy to determine what is required of us.

*****

I want to end with a pair of questions that have stuck with me since the first pages of the novel – who is anthroponomy for and what is its relationship to (in)justice? The author is fairly transparent about the former.[47] This is a novel for working through a particular sort of anxiety. But anthroponomy also purports to do more, to be *for* more, for it requires that we coordinate with others and, in so doing, learn to work through disagreement toward transformative change. *We need more tools like this.* Beyond breaking open worlds and sense, in disagreement we hold ourselves open and accountable to each other. This is a good principle/lesson for Ryders on the Farm and others, but who might feel less than safe holding themselves open to us? *And rightly so.*

Working through disagreement is not just a matter of trust. Or, rather, being committed to working through disagreement ought not require that trust always be symmetrical. "Good relationships must un-work and account for domination"[48] and this includes how to hold disagreement and create new common sense without exacerbating existing unjust vulnerabilities. In coordinating with those we have displaced, erased, neglected, and excluded, Ryders should expect to hold ourselves (relatively) more open and vulnerable and, thus, subject to deeper change than our partners. It means something different for the author to deny himself the option of simply dismissing the Anthropocene than it would to expect his Indigenous interlocutors to do the same.[49] Some differences are too toxic to work through or to engage with the same degree of openness on both sides. This need not preclude us from making sense and navigating disagreement together. On Ryder Farm and elsewhere, pursing anthroponomy under conditions of injustice must not only allow for but anticipate such asymmetries.

Overall, I believe that an anthroponomic orientation could do Ryders and the Farm a lot of good. As settlers grappling with our colonial legacy in the time of climate change, we are exactly the kind of place/family that anthroponomy is intended for. While it may not be the strategy that Indigenous peoples employ to hold colonial nations accountable, anthroponomy has the potential to be very useful for settler individuals and communities, like the Ryders, looking for ways to hold *themselves* accountable to others through the land. *That's okay.* Even here/ then it will need to be informed by the decolonial theories and practices of Indigenous peoples. *That's okay too. Wonderful even.* Anthroponomy need not be the first or the only tool that everyone works with for it to be a worthwhile innovation. For myself, I feel certain that anthroponomy will be one of the frameworks I use to inform my work on the Farm moving forward. And, like the author, I look forward to seeing what others make of and with it.

Julia D. Gibson is the Animal Ethics Postdoctoral Fellow in the Department of Philosophy at Queen's University. She/they works on palliative and remembrance ethics for the dead and the dying of climate change with the aid of Indigenous, Afrofuturist, and feminist science fiction fantasy narratives. She/they has authored publications in feminist bioethics, technology studies, environmental philosophy, and critical animal studies.

Notes

1 p. xvi.
2 In the sense Tuck and Yang employ. Eve Tuck and K. Wayne Yang, "Decolonization Is Not a Metaphor," *Decolonization: Indigeneity, Education & Society*, v. I, n. 1, 2012, pp. 1–40.
3 E.g., Greta Gaard, "What's the Story? Competing Narratives of Climate Change and Climate Justice," *Forum for World Literature Studies*, v. 6, 2014, pp. 272–291; Kate Rigby, *Dancing with Disaster: Environmental Histories, Narratives, and Ethics for Perilous Times*, Charlottesville, VA: University of Virginia Press, 2015; Donna Haraway, "Anthropocene, Capitalocene, Plantationocene, Chthulucene: Making Kin," *Environmental Humanities*, v. 6, 2015, pp. 159–165.
4 Robin Kimmerer reminds us that loving the land is not a purely internal affair confined to head and heart, but an active practice with real power. Robin Kimmerer, *Braiding Sweetgrass: Indigenous Wisdom, Scientific Knowledge and the Teachings of Plants*, Minneapolis, MN: Milkweed Editions, 2013.
5 Julia Gibson, "Climate Justice for the Dead and the Dying: Weaving Ethics of Palliation and Remembrance from Story and Practice," PhD Dissertation, Michigan State University, ProQuest Dissertations Publishing, 2019, 13898381; Julia Gibson and Kyle Whyte, "Philosophies of Science Fiction, Futures, and Visions of the Anthropocene," in Shannon Vallor, ed., *The Oxford Handbook of Philosophy of Technology*, forthcoming from Oxford University Press.
6 p. 135.
7 Kyle Powys Whyte, "Settler Colonialism, Ecology, & Environmental Injustice," *Environment & Society*, v. 9, 2018, pp. 129–144.
8 p. 9.
9 p. 11.
10 p. 56.
11 pp. 50–51.
12 Aaron Scott, "Cities," Radiolab, WNYC Studios, October 8, 2010, www.wnycstudios. org/podcasts/radiolab/episodes/91732-cities.
13 p. 35.
14 Though the Kantian undercurrents of anthroponomy do worry me. Kant's ethical and ontological frameworks have long been critiqued by feminists and decolonial scholars alike (e.g., Moyer, Mignolo). We'll need to keep an eye on those. Jenna Moyer, "Why Kant and Ecofeminism Don't Mix," *Hypatia*, v. 16, n. 3, 2001, pp. 79–97. Walter

Mignolo, "The Darker Side of the Enlightenment: A De-Colonial Reading of Kant's Geography," in Stuart Elden and Eduardo Mendieta, eds., *Reading Kant's Geography*, Albany, NY: SUNY Press, 2011, pp. 319–344.

15 SPACE on Ryder Farm (spaceonryderfarm.org) was co-founded by a cousin a few years back.

16 Known as *Platanus occidentalis* or, in Lenape, xaxakòk.

17 Ryder Farm Incorporated (RFI), "Animal Policy," 2019.

18 Ryder Farm Incorporated (RFI), "Animal Policy," 2019.

19 p. 154.

20 p. 29.

21 Tuck and Yang (2010, p. 35) write, "Reconciliation is concerned with questions of *what will decolonization look like? What will happen after abolition? What will be the consequences of decolonization for the settler?* Incommensurability acknowledges that these questions need not, and perhaps cannot, be answered in order for decolonization to exist as a framework. We want to say, first, that decolonization is not obliged to answer those questions – decolonization is not accountable to settlers, or settler futurity. Decolonization is accountable to Indigenous sovereignty and futurity."

22 What the Ryders think/feel we're up to is pertinent to the question of wantonness.

23 p. 132.

24 Kimmerer, 2013, p. 215.

25 Kimmerer, 2013, pp. 214–215.

26 pp. 85–86.

27 Kimmerer, 2013.

28 p. 131.

29 As a settler scholar, I feel very wary of making claims about the "core moral problem of coloniality" writ large (157). That the author's general critique of land abstraction is compatible with the work of many Indigenous scholars (e.g., Kimmerer, Maracle, Whyte) is reassuring. As for the deeper claim, I remain skeptically agnostic. Kimmerer, 2013; Lee Maracle, *Memory Serves: Oratories*, Edmonton, AB: NeWest Press, 2015; Whyte, 2018.

30 Tuck and Yang, 2012.

31 p. 139, endnote 107.

32 p. 41.

33 p. 10.

34 And, currently, we are.

35 Tuck and Yang, 2012.

36 p. 131.

37 p. 26.

38 p. 116.

39 pp. 51, 114.

40 p. 114.

41 Kyle Whyte, "Indigenous Science (Fiction) for the Anthropocene: Ancestral Dystopias and Fantasies of Climate Change Crises," *Environment and Planning E: Nature and Space*, v. 1, n. 1–2, 2018, pp. 224–242.

42 p. xv.

43 E.g., Tuck and Yang, 2012.

44 p. 111.

45 p. 87.

46 pp. 111–112.

47 p. xiii.

48 p. 58.

49 p. 104.

Glossary

I either *use* or *am informed by* the following words as defined, trying to be consistent. Sometimes along the book's way, however, there are variations on meaning depending on whether something is still being worked out at that point in the book or whether only a part of the meaning is the focus at that moment. Some of the following words are not explicitly used in the book but sum-up aspects of the book behind the scenes. Perhaps they will be illuminating for some. **Boldfaced** words in the definitions are themselves also defined here.

Accountability: (1) The process of acknowledging the moral claims of others, including what is implied for your behavior and attitudes in order to be consistent with that acknowledgment; (2) the capacity to engage in such a process

Anthropocene: (1) The proposed, geological name for the current, planetary time of Earth following the Holocene Epoch and the Meghalayan Age, and meaning, roughly, the epoch of the human; i.e., the time when Earth's geology is determined in significant ways by human life; (2) the geological age in which **anthroponomy** structures our social processes

Anthroponomy: (1) A species – a particular use – of **autonomy**; (2) a **teleological and mereological ordering** of autonomy to that particular use; (3) *the coordination of autonomy* within and across humankind through **fractal forms** *so that* humankind is collectively **accountable** for its effects on others across **planetary scales** of space and time within autonomous processes; (4) an **orientation** toward such coordination of autonomy, i.e., toward such a specific use of it

Anti-imperial/ism: (1) Resisting, taking apart, or finding alternatives to **imperialism** or imperial ways of life, including their organizational forms, character, motives, desires, fantasies, and standard manners of behavior; (2) an **orientation** to shape your life, community, or organization by (1)

Authenticity: (1) The process of maintaining true relationships, beginning with your relationship to yourself by which you might be said to be true to yourself; (2) owning your feelings, desires, fantasies, etc. and being true to your beliefs, intentions, and relationships; (3) evidence of such processes, sometimes understood as a quality; (4) (of "**moral authenticity**") the process of re-evaluating your moral relations

Autonomy (in-relationship):   (1) The process of finding out, living by; or creating practices, institutions, policies, laws, etc. that make **sense** to you, worked out in relationship with others, fundamentally through **disagreement** around the sense of the **world**; (2) that process focused especially around **accountability**, giving structure to sense in **community**; (3) the capacity to engage in (1)

Capitalism: (1) An economic form with the goal of generating maximal profit for private individuals and private agents (e.g., corporations, investment firms and their clientele, etc.) by locating all economic provisions in the market, that is, in transaction; (2) that same economic form using a system of abstract value that renders all goods and services fungible, making profit a matter of increasing one's transactional power; (3) the organizational, characterological, and behavioral forms, including mentalities, of societies intentionally supporting (1–2); the organizational, characterological and behavioral forms of societies, including mentalities enabling (1–3) by clearly having adapted to them

Colonialism: (1) The form of **imperialism** by which an **empire** – or a would-be one – aims to establish or maintain **colonies**; (2) the associated organizational, characterological, and behavioral forms, including mentalities, that directly support or indirectly adapt to enable (1); (3) the persistence of (2) even after **decolonization**

Coloniality: (1) The mentality and associated organizational, characterological, and behavioral forms by which **colonialism**, or aspects of it (e.g. racism, patriarchy), make **sense** to a given society; (2) that mentality as a precursor to **colonization**; (3) that mentality as a lingering presence of **colonialism** after **decolonization**

Colonization: The **imperial** process of establishing or maintaining **colonies**

Colony: A **land** and its beings (humans, animals, plants, etc.) settled by an **empire** for the sake of expanding it or providing it with more resources and opportunities, while maintaining the land's political and social status as peripheral to the empire

Community: (1) The process of living in common; (2) a group of people (or personified beings) living in common through a shared process

Decoloniality: (1) The mentality and associated organizational, characterological, and behavioral forms by which **coloniality** is resisted, taken apart or rejected; (2) a species of **anti-imperialism**; (3) a precursor to, part of, or legacy of **decolonization**

Decolonization: The process of freeing **colonies** from **empire**

Disagreement: The process of contesting the **sense** of the **world**

Domination: Protracted power over others in a way that does not make **sense** to them or to those who are morally **accountable** for them and in which no persistent attempt is made to resolve any **disagreement** surrounding that power

Empire: (1) A political form in which a given society seeks to expand its domain through **imperialism**; (2) the social form supporting, giving rise to, or longing for the political form, including mentalities and associated organizational, characterological, and behavioral forms

Fractal forms (of autonomy): (1) Forms by which the process of autonomy is organized according to different scales (e.g., spatial, temporal) in different domains (e.g., economic, social, political) with different scopes (i.e. those people concerned at the scale in the domain, such as "all people living in this **land**"); (2) such forms organized to be consistent with each other and to interconnect when appropriate, as, e.g., in **anthroponomy** and its **teleological and mereological ordering**

Heteronomy: (1) The condition of living by practices, institutions, policies, laws, etc. that do not make **sense** to you, are not worked out in relationship with others, or ignore or otherwise avoid **disagreement** around the sense of the **world**; (2) that condition denying especially **accountability**; (3) the lack of the capacity to engage in **autonomy**

Imperial/ism: (1) Beginning, maintaining, or protecting the agenda of expanding a given polity through **domination** of others and their **lands**; (2) an **orientation** to shape your life, community, or organization by (1)

Industrialism: (1) A system of production based historically in extracting resources from the Earth and in fossil fuel production, use, and admixture in other products that, historically, relies on Earth's capacities to absorb waste from both production and consumption; (2) a system of production (often that same system as in [1]) that aims to produce on a mass scale

Isonomy: (1) Equality of **accountability** in **sense**-finding and sense-making between all people in a **community**; (2) equality under the law, provided that the law makes sense; (3) norms between people who are **autonomous** and that do not produce **heteronomy**

Land: The field of myriad forms of life, all **morally considerable**, that sustains societies and is a condition for **communities**

Land abstraction: The extraction of **relational reason** from the **land** for the sake of its use within **colonialism, capitalism,** and **industrialism**

Land logic: The way in which **land** is conceptualized, for instance, through **land abstraction** or, oppositely, through **relational reason**

Landed autonomy: **Autonomy** involving **lands**

Mereological ordering: The way in which things are ordered as parts to a whole, for instance, as parts of **anthroponomy** respective to the whole task of it

Moral considerability: The quality of a being as deserving of some form of moral accountability

Orientation: A stable attitude, relational disposition, and practical stance toward something

Planetary scales (of space and time): The scales of space and time in which planetary, ecological, biochemical, or geological processes occur

Relational reason: A form of reason, cooperating with theoretical and practical reason but not reducible to either, in which familiarity with others, seeking to relate to others and to yourself, and being morally **accountable** to others structure an interpersonal process of being in **authentic** and **autonomous** relationships, including through **lands**

Self-determination: (1) The process of a person or of a group, including a political body, of living according to what makes sense to them **authentically**; (2) the capacity for such a process; (3) the **orientation** toward such a process

Sense: The meaning of something as it appears upon consideration

Social process obfuscation: Any way in which the social processes that form something (e.g., global warming) are obscured by the **sense** given to the **world**

Teleological ordering: The way in which intentional things are ordered to contribute to a greater purpose (e.g., moral **accountability** to humankind for the effects of global warming)

World: A dynamic, polysemic field of **sense**, including (1) the apparent continuum of common sense by which people live, often involving contradictions ("their/our world"); (2) the surrounding continuum of sense in which you live ("your world"); (3) that which disrupts a continuum of sense in which you or others live by showing that there is a wider field of sense of consider ("the world"); (4) the *locus* of the most intimate sense to you, subject then to critical reflection on it ("your world")

Figure E.2 Pleasant City, Ohio, 1938

Index

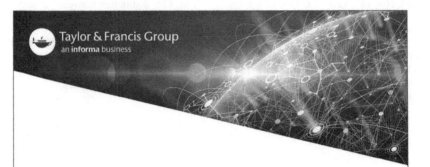